Low Invasion Coring

Best Practices Regarding Objectives, Planning, Equipment, Materials, Execution, and Follow-Up

J.B. Bloys
Chevron Energy Technology Company

H.R. Warner Jr.
Warner Consulting Services (formerly with ARCO)

Society of Petroleum Engineers

Printed in the United States of America.

The paper used in this book meets the minimum requirements of ANSI/NSIO Z39.48-1992 (R1997). ♾

ISBN 978-1-55563-303-5

11 12 13 14 15 16 17 / 9 8 7 6 5 4 3 2 1

Society of Petroleum Engineers
222 Palisades Creek Drive
Richardson, TX 75080-2040 USA

http://www.spe.org/store
books@spe.org
1.972.952.9393

Low Invasion Coring

Foreword

This SPE monograph has been developed to assist the reader in understanding how to apply the array of technologies required for the successful completion of a low-invasion coring (LIC) project.

The purpose of LIC is to cut cores that have a significant portion of the core cross section uninvaded by mud filtrate. To maintain the value of the uninvaded rock, one must use proper core-handling and preservation protocols, and then perform high-quality routine-core-analysis-laboratory (RCAL) and special-core-analysis-laboratory (SCAL) tests on these portions of the rock. The overall goal of this effort is to obtain more-accurate rock property data that can then be used for hydrocarbons-in-place, reserves, and reservoir-development calculations.

The monograph is structured both to discuss the principles of obtaining low-invasion cores and provide practical information to engineer field LIC projects. The body of this monograph discusses these matters in fairly specific terms and also gives in-depth background information. The appendices provide detailed information on specific core-analysis procedures.

Note: The discussion of specific equipment from various manufacturers and service companies is up to date as of late 2010, but the particulars may become somewhat out of date in future years. While the LIC principles remain constant, the specific manufacturers and service companies should be contacted concerning the newest equipment for LIC projects.

Acknowledgments

This monograph would not be possible without the very generous help and thoughtful insights of the subsequently listed people. It is a far better document because of the inclusion of their exceptional knowledge and experience. Special thanks are extended to Jim Hedges for his extraordinary work as technical editor.

This work is heavily based on the extensive and detailed technical work by Jake Rathmell during his years at ARCO. His innovation and persistence made the goal of LIC a reality. Words alone cannot adequately express our appreciation and thanks to Dr. Rathmell.

Jake Rathmell, ARCO, retired
Jim Hedges, ConocoPhillips Petroleum Company
Ed deZabala, Chevron Energy Technology Company
Luc van Puymbroeck, Baker Hughes
Tom Fate, Chevron Energy Technology Company
Kevan Sincock, BP
Steve Polnaszek, Texaco, retired
Jim Headley, Texaco, retired
Frank McCaffery, Chevron Energy Technology Company, retired
James Parr, Cabot Oil & Gas Corporation
Melanie Dunn, Omni Laboratories
Bob Skopec, Petrophysical Applications International
Dave Rodman, Halliburton
Gordon Tibbitts, ProDril

Contents

Foreword v
Acknowledgments vi

1. Overview 1
- 1.1 Introduction 1
- 1.2 Objectives of Low-Invasion Coring 3
- 1.3 Key Aspects of Low-Invasion-Coring Technologies 4
- 1.4 Historical Perspective on Low-Invasion Coring 5
- 1.5 Additional Insights 5
- 1.6 Added Benefits 6
- 1.7 Limitations of LIC 6
- 1.8 Topics Covered in This Monograph 7
- 1.9 Terminology and Styles Used in This Monograph 8

2. Preplanning 9
- 2.1 Clear Objectives 9
- 2.2 Team Approach 10
- 2.3 Project Manager 11
- 2.4 Adequate Lead Time 11
- 2.5 Coring and Core-Analysis Procedures Document 12
- 2.6 Specialized Personnel at Rigsite and Laboratory 14

3. Low-Invasion Coring Bit Selection 15
- 3.1 General Features 15
- 3.2 Bit Selection vs. Rock Type 17

4. Other Downhole Hardware 19
- 4.1 Inner Core Barrels 19
- 4.2 High-Torque, Outer Core Barrels 19
- 4.3 Mud Motors 20
- 4.4 Core-Catcher Assemblies 20

5. Low-Invasion Coring Fluids 21
- 5.1 Overview 21
- 5.2 Water-Based Coring Fluids 22
- 5.3 Oil-Based Coring Fluids 25
- 5.4 Mud-Filtrate Tracers 29
- 5.5 Mud Displacements 33
- 5.6 Solids Control 35

6. Rigsite Operations 37
- 6.1 Overview 37
- 6.2 Personnel 37

6.3 Coring Fluids—Displacement, Tracers, and Sampling 39
6.4 Equipment—Coring Contractor and Core-Analysis Contractor 39
6.5 Strapping and Drifting the Pipe 45
6.6 Making Up the Core Barrel 45
6.7 Tripping In 46
6.8 Coring Mechanics 46
6.9 Tripping Out 47
6.10 Core Lay-Down 47
6.11 Core Processing 49
6.12 Quality-Control Sampling 50
6.13 Shipping to Core-Analysis Laboratory 50

7. Coring Poorly Consolidated Sandstone 53
7.1 Overview 53
7.2 Core Length and Diameter 53
7.3 Coring Fluids 53
7.4 Mud Overbalance Pressure 54
7.5 Bit Selection and ROP 54
7.6 Flow Rate and Rotary Speed 54
7.7 Full-Closure Core Barrel 55
7.8 Tripping Rates 55
7.9 Core Stabilization 56
7.10 Personnel 57

8. Core Analysis 59
8.1 Operator Representative at Laboratory 59
8.2 CT 59
8.3 Core Spectral Gamma Ray Logging 59
8.4 Core Slabbing, Plugging, and Preservation Protocols 59
8.5 Core-Analysis Procedures 66
8.6 Poorly Consolidated Cores 69
8.7 Contacts 71

Nomenclature 73

Glossary 75

References 77

Appendix A—Selected SPE Papers Concerning LIC Technologies 81

Appendix B—Special Circumstances—Sponge Coring and the Use of Noninvading Core Gel 85
B.1 Gel Coring 85
B.2 Sponge Coring 85

Appendix C—Calculation of Filtrate Saturation 89
C.1 Definition of Terms 89
C.2 Analysis Procedure 89

Appendix D—Calcium Carbonate Determination in Whole Mud **91**

Appendix E—Iodide Ion Analysis **93**

Appendix F—Bromonaphthalene-Analysis Procedure **95**
F.1 Introduction 95
F.2 Scope 95
F.3 Instrumentation 96
F.4 Reagents 96
F.5 GC Conditions 96
F.6 Analytical Procedure 96
F.7 Tracer Calculations 97
F.8 Definition of Terms 98

Appendix G—Sample Anaerobic-Crude-Oil Collection Letter **99**

Appendix H—Chloride Ion Analysis—Brine Saturation **101**
H.1 Procedure 101
H.2 Definition of Terms 102

Appendix I—Impact of Oil-Based Mud on Connate Water Saturation and Other Rock Properties ... **103**
I.1 Background 103
I.2 Relationships of Mud-Filtrate/Connate Brine IFT and Original Reservoir Properties ... 103
I.3 Summary 111

Appendix J—Deuterium Oxide Tracer Calculations **113**

Author Index **115**

Subject Index **117**

Chapter 1

Overview

1.1 Introduction

Low-invasion coring (LIC) is a term used to describe the technologies and procedures for obtaining core samples from oil and gas reservoirs as undisturbed as possible by the coring and core-handling processes. With standard coring, mud filtrate almost always invades to the center of the core, preventing or lessening the certainty of accurate measurements of, for example, water saturation (S_w), wettability, and formation brine composition. Therefore, LIC seeks to acquire 4-in. diameter cores with at least the center 1.5-in. diameter cross section uninvaded by mud filtrate. Cores taken in this manner ensure (a) more-reliable connate S_w values and other routine core analysis (RCAL) data and (b) more-reliable core samples for various special core analysis (SCAL) measurements. These measurements are critical to the accurate determination of original in-place hydrocarbon volumes and reserves, as well as for the calibration of logs. Even somewhat unexpected phenomena like vertical formation-brine salinity variations become more apparent when uninvaded cores are obtained and analyzed. **Figs. 1.1 through 1.3** illustrate core samples with various degrees of invasion.

This document provides engineers, geologists, and drillers with an understanding of the technologies and procedures to be used to successfully complete an LIC operation. The overall goal for this document is to provide enough detailed information for technical personnel to properly design and conduct an LIC project. Examples are provided of specific procedures, and insights are given on each contributing technology; therefore, the process can be adapted to a variety of coring situations.

Note: The document covers coring and core-handling equipment, as well as materials, technologies, and procedures. However, the RCAL and SCAL laboratory-measurement procedures are only discussed in limited detail herein (see Chapter 8). Other lengthy documents have been prepared by industry experts covering those topics—particularly the American Petroleum Institute (API) *RP40 Recommended Practices for Core Analysis* (2nd ed., 1998). Also, commercial laboratories typically have standard procedures to follow for these measurements, and/or the contracting company's technical specialists have their own preferred procedures for running such tests.

There is a range of special measurements enabled by LIC technology, from chloride variations in intervals at connate S_w that do not produce water, to native-state testing with both water-based-mud (WBM) and oil-based-mud (OBM) cores. None of these special measurements are standard with commercial laboratories. Generally, the surest way to not achieve complete success in an LIC project is to leave testing procedures details strictly to the commercial laboratories' standard practices.

Fig. 1.1—White light photograph of uninvaded core center, 4-in. core cut with early asphalt oil-based coring fluid. From Woodhouse (1998).

Fig. 1.2—Uninvaded core center from North Sea gas field, using water-based mud ($\varphi \approx 10\%$ BV, $k \approx$ 0.5 md, $S_w \approx 35\%$ PV). From Woodhouse (1998).

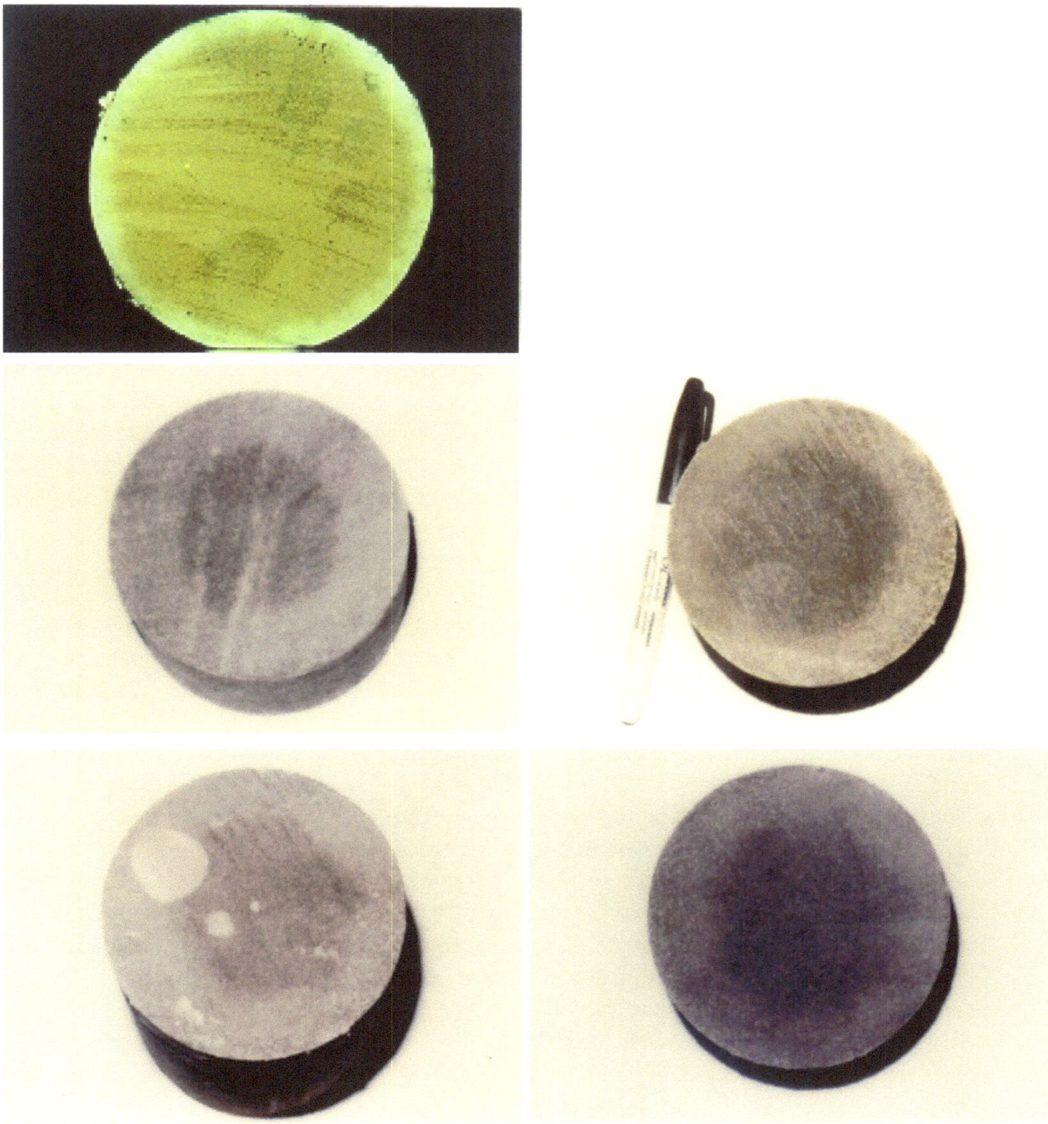

Fig. 1.3—Alaska North Slope partially uninvaded cores. Top—UV photo of 4-in. core with only ¼-in. oil-mud-filtrate invasion; the remaining photographs show early water-based mud cores with moderately uninvaded portions.

1.2 Objectives of Low-Invasion Coring

There are two primary objectives of LIC technology:

1. Obtain uninvaded core samples for RCAL Dean-Stark water volume measurements (in zones where the water phase is immobile), and other routine measurements, such as porosity, permeability and connate water salinity determinations.
2. Obtain uninvaded core samples for SCAL measurements, such as electrical properties, capillary pressure, relative permeability, and wettability.

While it is easy to state these objectives, each objective must be understood regarding all the effort and detailed activities required to achieve these objectives in a particular project. Also, there are limitations to each objective, as discussed in Section 1.7.

1.3 Key Aspects of Low-Invasion-Coring Technologies

To achieve LIC core samples, the following combination of technologies must be applied. Many of these technologies concern the core bit, core catcher, and core barrel. **Fig. 1.4** shows the physical relationships of these mechanical parts of the coring system. The core barrel consists of an outer and inner core barrel. The outer core barrel is made of steel. The inner core barrel is typically made of aluminum, but for some applications, either fiberglass or steel is used.

- **Core Bits.** Specialized bits have been developed to core as fast as possible, and to direct the coring fluids away from the core. The advent of polycrystalline-diamond-compact (PDC) bits has allowed cores to be cut at coring rates up to 200 ft/hr and averaging more than 60 ft/hr in high quality sandstone intervals. Before PDC bits, cores were typically cut at coring rates significantly less than 10 ft/hr. At these low coring rates, even with the use of the best available drilling muds, mud filtrate usually invaded to the center of a 4-in. diameter core.
- **Coring Fluids.** Simply coring at a high rate of penetration is not enough to ensure LIC core samples. The drilling fluids must be designed to minimize the mud-filtrate loss (i.e., mud-filtrate invasion into the core) during the coring operations. Also, the amount of overbalance pressure influences the extent of mud-filtrate invasion into the core. Greater overbalance gives a larger pressure differential for fluid invasion, but this greater driving force is partially offset by the lower permeability of the compressible filter cake, especially with WBM. Tracers should be added to the mud system to track how far mud filtrate has invaded into the core cross section.
- **Core Barrel.** It is important to minimize exposure of the core to mud-filtrate invasion downhole after the core is cut. The core continues to be exposed to additional mud-filtrate invasion as it moves up the throat of the coring bit and into the core barrel.
- **Core Plugging.** To prevent diffusion of the mud filtrate from contaminating the interior of the core, the core should be slabbed and plugged as soon as possible after the core reaches the surface. The speed with which this procedure can be accomplished is dependent on the particular circumstances at the rigsite, but the objective is the same in all cases—cut the individual core plugs as quickly as possible. Included in this step is the use of core sample-preservation procedures and materials that do not alter the properties of the core samples, such as deoxygenated mineral oils or stabilized field crude oil.
- **Core Diameter.** It is best to cut as large a diameter core as possible. The discussion presented in this document focuses on cutting 4-in. diameter cores. The 5.25-in. diameter cores can seem

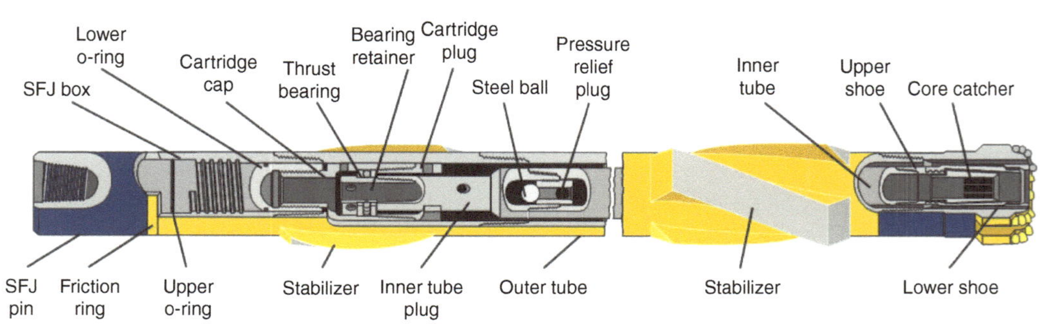

Fig. 1.4—Cutaway diagram of core barrel (used with permission from Baker Hughes).

attractive; but because they are cut with a 12.25-in. bit, the rate of penetration (ROP) is usually too slow for reliable LIC. However, in soft rock, this option may be preferred. Conversely, with a 2.75-in. diameter core, there is likely to be a very limited uninvaded center portion of the core because of its small diameter.

LIC projects are not simple, but when executed correctly, the higher-quality data significantly improves reservoir characterization, which is one of the largest risk factors in any field development.

1.4 Historical Perspective on Low-Invasion Coring

With typical applications of historic coring technology—even when using a good-quality mud—mud filtrate invaded to the center of the cores of any reservoir rock with reasonably high permeability [i.e., a few millidarcies (md) or more]. Jenks et al. (1968) published a set of laboratory microbit measurements in which the effect of WBM filtrate on residual oil saturations was studied. They concluded that overbalance pressure was the prime factor affecting the degree of oil stripping, with mud composition and penetration rate as other important variables.

In the late 1970s, as PDC bit technology was being developed, the possibility of developing a "low-invasion core bit" was discussed by Huff et al. (1978). These authors recognized that a combination of rapid coring rate and high-quality drilling mud was needed to achieve the desired results.

In the 1980s, the LIC concepts were proved and became a reality. A series of papers were published about technical studies related to the coring and core analysis in the early 1980s of the Sadlerochit reservoir in the Prudhoe Bay field. Two of these papers included a photograph of the invaded core centers of 4-in. diameter cores cut in the Sadlerochit gas-cap interval; see Fig. 1.1 (Holstein and Warner 1994; Woodhouse 1998). Another paper presented detailed microbit laboratory coring tests results using conventional diamond bit technology, which indicated that at higher coring rates "one could expect the volume of filtrate invasion to be even less, and for partially uninvaded cores to result" (Warner and Rathmell 1997). A fourth paper concerning OBM 2.75-in. diameter field pressure coring found that with the coring rate averaging approximately 30 ft/hr (using PDC coring bits), the amount of filtrate invasion was only approximately 0.7 pore volumes (PV) (Charlson et al. 1997).

During the 1990s, several papers were published that described further advances in LIC technology. A few of these papers that discuss some aspects of LIC technology development and its application are Bloys et al. (1993); Wylie et al. (1993); Rathmell et al. (1995, 1999); Tibbitts et al. (1990); Egbogah et al. (1997); McCoy et al. (1997); and Dawe and Murdock (1990).

These technical papers show that during the period of 1975 through 2000, LIC went from a concept to a fully developed set of technologies now being applied worldwide. As also shown by these papers, each application of LIC technology requires a combination of the accumulation of the knowledge and experience gained from past studies and applications, and the recognition of what will be new and different in the next proposed application.

1.5 Additional Insights

Careful preplanning is needed to assemble the right personnel, bits, coring fluid, coring equipment, and procedures. Everything is focused on limiting filtrate invasion by cutting the core as quickly as possible and tightly sealing the core during preservation. Cautious tripping procedures and careful surface-handling procedures reduce post-coring damage or rock/fluid changes.

Once the core is successfully cut with low mud-filtrate invasion, care must be taken to preserve the uninvaded portions of the rock. When the core reaches the surface, it must be quickly sealed in an airtight container to prevent any water evaporation and oxidation of exposed rock surfaces. Also, because the mud filtrate in the outer edge of the core diffuses toward the core center, the core must be very quickly slabbed and plugged, and the core plugs preserved individually. This procedure can be done either on site or in a nearby laboratory if the work can be completed within 48 hours of the core reaching the surface. Tracers should be included in the coring fluid for the mud-filtrate invasion to be tracked and quantified. Quality control of the core-analysis process is critical to avoid water loss, cross-contamination, and systematic errors.

LIC projects are more expensive than standard coring projects (e.g., requiring extra personnel and transportation costs, better muds and tracers, more core tests) and are not appropriate for every coring project. The best approach is to obtain an LIC early in the life of a major field (i.e., the second or third well) for in-place hydrocarbon volumes and reserves to be more accurately determined, to have more-accurate log calibrations applied to all subsequent noncored wells, and to have more-reliable numerical simulation models from which to plan the reservoir development.

LIC projects are quite complex, and the desired higher-quality core data might be compromised at a number of critical junctures. Using a veteran coring project manager—familiar with all facets of the technology and capable of overseeing the project planning, fieldwork, and laboratory analyses—substantially increases the probability of success. It is important that the coring project manager carefully coordinate the input and requirements of the project geologist, reservoir engineer, drilling engineer, rig supervisor, materials/logistics department, drilling fluids company, coring company, and commercial core-analysis laboratory.

Finally, if accurate water saturations with a maximum degree of certainty are the primary objective of an LIC project, then an OBM with low water content must be used as the coring fluid. This approach avoids all possible sources of water contamination of the connate brine in the reservoir intervals.

1.6 Added Benefits

Designing the coring fluid to preserve the natural in-situ state of the core also makes the fluid likely to cause very little formation damage to the rock matrix. Well tests after LIC usually show very little formation damage, which is consistent with the fact that LIC fluids are very similar to low-damage, drill-in fluids designed for openhole completions, although the filter cake from an LIC fluid is not as easy to remove.

The same approaches that minimize formation damage also increase borehole stability, which decreases rig time and improves core recovery.

The high ROP produced by LIC bits can significantly reduce the rig time necessary to cut the cores. ROPs of <5 ft/hr were common with old-style natural diamond bits, but coring rates of 60 ft/hr (1 ft/min) or more are common with low-invasion PDC bits. In fact, low-invasion style bits are frequently chosen for standard coring projects simply because of the high ROP.

Because LIC fluids produce very high-quality filter cakes, stuck pipe is uncommon, even with high overbalance pressures. The extra project-planning and well-design considerations for an LIC project can minimize core-barrel jamming while increasing core recovery and coring efficiency.

1.7 Limitations of LIC

The ideal scenario is to obtain pristine, unaltered reservoir rock samples for subsequent laboratory testing. However, this ideal scenario cannot be achieved for a variety of reasons in either conventional coring or LIC. To avoid having unreasonable expectations, these limitations must be discussed. A totally pristine reservoir rock sample should have the following:

- Fluid content (i.e., connate water, oil, and gas) and its distribution within the pore system unaltered by the coring and core-handling operations
- Gas composition and gas-in-solution in the connate water and oil unaltered by the coring and core-handling operations
- pH of the connate water and eH (i.e., oxidizing/reducing state) within the pore system unaltered by the coring and core-handling operations
- Pore pressure and multidirectional stresses on the rock at reservoir conditions unaltered by the coring and core-handling operations
- Temperature maintained at that of the reservoir during the coring and core-handling operations

LIC technologies can be used to keep most of the immobile fluid content and its distribution within the pore system—particularly the connate water in the hydrocarbon zones—essentially unaltered by the coring and core-handling operations. However, the temperature and stress on the reservoir rock cannot be maintained at reservoir conditions because the coring process alters these properties immediately as the core is being cut. The pore-pressure, gas-composition, pH, and eH aspects are altered, unless a pressure core barrel is used to retain downhole pressure conditions and keep the gas in solution.

LIC technologies have been used successfully for the purpose of obtaining core samples at connate S_w and its original S_w distribution in a fairly wide range of rock types, reservoir temperatures, and hole angles. Situations in which complications develop are as follows:

- **Presence of Mobile Water.** When mobile water is present, such as below an oil/water contact (OWC) and in the oil/water transition zone, the in-situ S_w is reduced by any mud-filtrate invasion and/or by the bleeding of fluids from the core during the depressurization of the core from reservoir to surface conditions. This reduction also applies to the oil column interval in a waterflooded reservoir.
- **Countercurrent Imbibition.** In gas reservoirs, countercurrent imbibition moves either WBM filtrate or especially OBM filtrate to the core center—even if the initial mud-filtrate invasion only impacts the outer inch of the 4-in. diameter core. This movement occurs because gas is the least wetting of the various fluids. However, this mechanism does not alter S_w if a properly designed OBM is used. In oil reservoirs, countercurrent imbibition is a much slower process, and any immediate saturation change is likely to be insignificant. In all cases, the way to minimize this effect is to separate the center core samples from the rest of the core as quickly as possible once the core is available for processing at the surface. The extent of this effect, if any, can be quantified by the use of tracers in the mud to track the movement of the mud filtrate and its depth of invasion.
- **Poorly Consolidated Rock.** High core recovery is difficult any time poorly consolidated rock is cored. Chapter 7 presents guidance about the best procedures for applying LIC technologies for use with this rock type.
- **Very Hard Rocks.** Highly cemented formations, such as those with significant amounts of either quartz or quartzite overgrowths—or carbonate formations—do not core at rates of more than a few feet per hour with any bit type and are much more likely to be invaded with mud filtrate to the core center. Accurate water saturations can be obtained with a properly designed OBM—given that the connate S_w is not mobile. Because of the surfactants in the filtrate, it is unlikely that the original rock wettability can be fully maintained with OBM. If wettability is a critical goal, it can be maintained by use of a bland (i.e., low-pH, no surfactants) WBM.
- **Heavy, High-Viscosity Oils.** Extensive fingering/invasion of low-viscosity, WBM filtrate can occur because the displaceable pore volume is low because of the high mobility of the water phase relative to the low mobility of the oil phase. A small amount of filtrate can invade a substantial distance toward the center of the core.
- **Low-Displaceable Pore Volume.** Another case with low displaceable pore volume is rock with high connate S_w mainly filling the microporosity. As previously discussed, a small amount of filtrate penetrates farther than expected toward the center of the core because the displaceable pore volume is low.

1.8 Topics Covered in This Monograph

To maximize the desired results from applying LIC technologies, a very high-quality drilling mud must be used, the core must be cut as rapidly as possible, core surfacing and handling time must be minimized, and core samples must be removed from the whole core as soon as possible and preserved properly for later laboratory testing. The rest of this document presents details of these various technologies that fall under the umbrella of "LIC technologies."

- Chapter 2—Preplanning
- Chapter 3—Low-Invasion Coring Bit Selection
- Chapter 4—Other Downhole Hardware
- Chapter 5—Low-Invasion Coring Fluids
- Chapter 6—Rigsite Operations
- Chapter 7—Coring Poorly Consolidated Sandstone
- Chapter 8—Core Analysis
- Appendix A—A listing of SPE papers discussing the development and applications of LIC technologies, and related papers from other technical journals.
- Appendix B—Special Circumstances—Sponge Coring and the Use of Noninvading Gel
- Appendix C—Calculation of Filtrate Saturation
- Appendix D—Covers a method for determining the amount of calcium carbonate in a WBM
- Appendices E and F—Describe the particulars of various laboratory-analysis procedures for tracer analysis
- Appendix G—A sample letter from an Operating Company Technology Center to Field Personnel for the collection of anaerobic crude-oil samples
- Appendix H—Covers the analysis of formation brines by determining the chloride ion concentrations
- Appendix I—Discusses S_{wc} vs. interfacial tension (IFT) and pressure gradient
- Appendix J—Deuterium oxide tracer calculations

1.9 Terminology and Styles Used in This Monograph

The following terms are used in this monograph, but several similar terms have the same meaning:

- The term "rigsite" is used throughout, but equivalent terms are "rig" and "wellsite."
- The terms "drilling mud" and "coring fluid" (or similar terms) are interchangeable.
- In most places, numbers are stated as decimal numbers; that is, 1½ is stated as 1.5, or 12¼ is stated as 12.25, and are expressed in English units of inches.

Chapter 2

Preplanning

2.1 Clear Objectives

The first steps in any project as complex as a low-invasion coring (LIC) project are to clearly establish goals for the project and establish an adequate budget. Once the goals are clearly defined, the project can be designed to give the highest probability of acquiring the desired data. The coring project manager should work with the project geologists and reservoir engineers to determine the need for, and priority of, the various types of core data.

2.1.1 Questions To Ask About the Need for Low-Invasion Coring (LIC).

- Is an LIC truly needed, or are standard porosity, permeability, and lithology data from a conventional core sufficient?
- Are highly accurate water-saturation data needed for determination of original hydrocarbons in place and reserves?
- Is precise log calibration (including connate water salinity) needed for analyzing future non-cored wells?
- Are relative permeability data needed for the selection and quantification of the best reservoir-development approach?
- Are the core data to be used for reservoir modeling?
- Are the core data to be the basis for equity-allocation calculations?

Once these questions are addressed, the scope of the project can focus on the key issues.

LIC costs more (as does any coring seeking accurate connate water saturations) than standard coring, because there are more people, materials, equipment, time, and analyses involved. Everyone involved should understand and accept this increased cost from the start. Not every coring project needs to pursue low mud-filtrate invasion, but if the data required from the core can only be reliably gathered from uninvaded rock, then the extra expense is well justified.

Trying to obtain a low-invasion core from an exploration well can be very frustrating because coring depths are difficult to identify and rock types are not yet known, which affects bit choice. However, the second or third delineation well following field discovery is probably a good candidate for LIC. The resulting data are available early in the life of the field when they can be used to define an optimal development plan. Low-invasion data from one well (maybe two wells) are often sufficient for a given field. Using these data for careful log calibration can make coring subsequent wells unnecessary. However, resolving equity-allocation issues with field partners may require cores from multiple areas of a field. *Note: An LIC using WBMs may also be used to obtain measurements of residual oil saturation, to evaluate waterflood recovery, or to evaluate oil saturations and areal/vertical conformance for miscible gas projects.*

Do not attempt *oriented coring* during an LIC project. The short knives that scratch the core to record orientation increase filtrate invasion by scraping off filter cake and increasing the probability of core jamming.

2.2 Team Approach

For an LIC project to be successful, input from, and the cooperation of, many different people with various technical specialties is needed. Assembling an appropriate team early in the project and keeping everyone fully informed at each stage of the project can greatly increase the probability of success. The team should include at least the following:

- Coring project manager
- Drilling department representative
- Geology department representative
- Reservoir engineering representative
- Coring contractor
- Drilling-fluid company
- Rig supervisor
- Commercial core-analysis-laboratory representative
- Materials/logistics department

Team members may have slightly different agendas for the well, but the earlier the team meets, learns the details of the well, and agrees to a set of priorities, the lower the chance that a problem can subsequently lead to project failure. At the start of the LIC project, all the information needed/desired for the project plan is typically not available; but if everyone has as much information as possible as early as possible, fewer incorrect assumptions are made that lead to complications in completing the project.

The early stages of the project are the time for asking the "thousand questions," as follows:

- What are the project goals?
- Do we have the necessary management support and budget?
- Are there any data from nearby wells? Logs? Mud logs? Cores? Coring logs? Bit records? Mud records? Rock mechanics data? ROP logs? Pore-level rock data? Pore-pressure or fracture-gradient data? Formation temperature? Formation brine composition?
- What rock types are expected? Sandstone? Limestone? Fractures? Poorly consolidated? Hard and heavily cemented? Sand/shale sequences? Shale types?
- Is either oil or gas expected? Both?
- What is the expected range of reservoir pressures?
- Is the preliminary well plan available yet?
- What is the casing program?
- What are the hole size, hole angle, and depth at core point?
- Are multiple zones to be cored? How far apart are the intervals to be cored?
- Are there any drilling problems expected? Lost circulation? Fractured rock? Salt layers? H_2S? Unstable shales? Over pressured zones? Other?
- What rig is to be used?
- Does the rig have a top drive?
- What is the layout of the mud system?
- How experienced is the rig crew?
- What problems can occur when displacing to the coring fluid?

- Is there adequate space available to process the cores at the rigsite?
- Can some rig hands be made available to help with handling and processing the cores?
- Are there mud loggers on the well?
- Are there sleeping quarters at the rig for the coring project manager, coring engineers, core laboratory technicians, or others?
- Where is the rigsite?
- How complicated are the rigsite logistics considerations?
- What are the lead times needed for equipment and materials?
- Is it unusually hot, cold, wet or dry—desert, Arctic, or rain forest?
- Is there a commercial core-analysis laboratory nearby?
- Does the commercial core-analysis laboratory need to be upgraded for this project?
- Can the core be sent by a courier service to an appropriate laboratory?
- Does the core have to be plugged and preserved on site?
- Are there any government regulation issues (i.e., customs, data/rock transfer restrictions) if national borders are crossed in transporting the cores or core plugs?
- Who is the mud contractor? The coring contractor? The commercial core-analysis laboratory?
- Can key, experienced people from each organization be assigned to the project?
- Can the coring project manager be allowed to determine/influence the choice of the contractors hired for this project?

2.3 Project Manager

As with any complex undertaking, the right project manager is essential for a smooth, successful LIC project. The ideal project manager needs, in addition to good organizational and interpersonal skills, some knowledge in drilling-rig operations, coring-bit selection, coring mechanics, drilling fluids, core processing/preservation, core analysis, and reservoir engineering. It is quite rare to find someone with this wide range of technical expertise and even rarer to find someone willing to spend the necessary amount of time at the rigsite performing the required project oversight. The alternative is for the project manager to develop a network of trusted experts to call while continuing to coordinate the activities of the various technical specialists. The project manager must have a strong personality and good communications skills to achieve the key goals of the project when faced with the inevitable pressures to minimize costs, minimize on-site personnel, and expedite operations. In most cases, the project manager has coring and/or core-analysis experience from within the technology branch of the operating company.

2.4 Adequate Lead Time

LIC projects cannot be planned and executed in just a few weeks. If at all possible, the preferred planning timeframe is 4 to 6 months, although this time can be reduced in some circumstances if the logistics are well-developed and a routine project is undertaken. The long lead time is needed to complete the following:

- Develop and refine the well plan, influence the choice of coring and core-analysis companies, and develop a realistic budget for all associated activities.
- Gather the data necessary for bit selection (i.e., records review, rock mechanics tests on cores from nearby wells, and review of logs.).
- Coring fluid pilot tests, spurt-loss tests, and IFT measurements (if OBM is used).
- Execute contracts with the coring contractor and the commercial core-analysis laboratory.
- Make any required upgrades to the local laboratory of the core-analysis company.
- Arrange for preferred personnel to be at the rigsite.
- Prepare a detailed coring and core-analysis procedures document.
- Manufacture the coring bits.

- Import and/or transport the equipment and materials.
- Arrange logistics either for a rapid core transfer to a laboratory or for processing on site.
- Order the tracers, if not readily available.
- Interface with the rig supervisor on rig-specific issues.

2.5 Coring and Core-Analysis Procedures Document

An LIC/core analysis is a complicated process requiring many months of activities, involving numerous contracted companies and individuals, and involving a wide array of disciplines, equipment, and materials. A detailed procedures document is needed to keep everyone aware of their roles and the critical paths to project success.

Coring and core-analysis procedures documents can be adjusted for the requirements of a given project, but a reasonable outline is included in the following subsections.

2.5.1 Coring and Core-Analysis Procedures Document Outline.

1. Coring Objectives—any or all of the following:
    - Lithology (geologic description)
    - Porosity
    - Permeability
    - Water saturation (for original-hydrocarbons-in-place and reserves estimates)
    - Brine composition/resistivity as a function of depth (log calibration)
    - Water- or gasflood conformance, recovery, oil saturation
    - Formation-damage potential
    - SCAL tests (i.e., relative permeability, electrical properties)

2. Project Overview
    - Broad outline of an LIC process
    - Project timing and time line
    - General well details
    - Geographic location
    - Special logistics issues
    - Listing of key service companies

3. Rigsite Operations
    - Safety
    - Well plan
    - Personnel—project manager, two coring engineers, mud specialist, two to four core-analysis technicians
    - Procedure for picking core points and core-barrel lengths
    - Procedure for acquiring quality-control/quality-assurance (QC/QA) mud samples
    - Primary, secondary, and spare coring bits: selection criteria [e.g., number of blades, total flow area (TFA), cutter type, back-rake specifications]
    - Core catcher shoes and outer/inner core-barrel specifications
    - Inner-barrel/bit-spacing specification
    - Rigsite equipment lists
        - Coring contractor
        - Commercial core-analysis laboratory
    - Coring fluid
        - Formulation

    - Properties specifications
    - Displacement
    - Maintenance
    - Tracer(s) and mud sampling
    - Solids control
- Plan to "strap out the drillpipe" (i.e., measure its length) for the depth of the top of each core to be accurately determined
- Plan to drift the pipe to be sure of the largest drop-ball passing
- Plan to include a drop-ball circulating sub to allow the circulation of bottoms up without flowing coring fluid through the core barrel and to place lost-circulation material (LCM) if required
- Guidelines for weight on bit (WOB), drillpipe revolutions per minute (rev/min), ROP, and mud-flow rate
- Trip rates to avoid gas expansion damage to the cores during surfacing
- Procedures for wellsite core gamma ray scanning with portable unit if appropriate
- Procedures for core lay down, marking, cutting, sampling, purging, capping, and transportation

4. Core-Analysis Procedures
    - Operator representative in laboratory for QC/QA
    - Computer tomography (CT) scanning before slabbing (especially with poorly consolidated rock)
    - Standard gamma ray
    - Spectral gamma ray (preferred)
    - Slabbing and plugging protocols (i.e., methods and frequency), including plugging fluids for RCAL and SCAL samples
    - Core preservation
    - Permeability to oil at connate water saturation (k_o at S_{cw})
    - Dean-Stark water-volume extraction of core plugs
    - Core plug cleaning and drying
    - Chloride (i.e., formation brine) analyses
    - Porosity, air-permeability, and grain-density measurements
    - SCAL electrical properties and capillary pressure measurements
    - Relative permeability measurements
    - Tracer-analysis procedures
    - Photography [i.e., ultraviolet (UV) and white light]
    - QC
    - Destination/distribution of data and core material
    - Thin sections, X-ray diffraction (XRD), scanning electron microscopy (SEM), mercury pore-size distribution, and any other desired measurements
    - Special circumstances
        - Examples: Freeze protection in the Arctic or air freighting core from remote locations

5. Contingency Plans
    - Preapproval for additional core runs if needed because of core jamming, poor choice of initial core point, or other complicating factors
    - Backup plan for transporting core from rigsite to off-site facilities
    - Spare bits

6. Contact List
    - Phone, mobile phone, fax, email, address for all LIC project participants

2.6 Specialized Personnel at Rigsite and Laboratory

Preplanning is crucial to an LIC project, but not sufficient. Careful oversight is required at the rigsite and at the core-analysis laboratory. Several LIC projects have been unsuccessful because of a lack of experienced personnel providing oversight. Invariably, situations arise that are not specifically covered in the coring and core-analysis procedures document, and then nonoptimal decisions might be made. Only a limited number of people are experienced in LIC operations, which can lead to problems when multiple projects are occurring simultaneously. Pairing experienced and inexperienced people is one way to increase the number of people who can apply these technologies. The specialized personnel include

- **Project Manager.** This person must be familiar with all aspects of the LIC project.
- **(Two) Coring Engineers.** Besides having strong general experience, both engineers must understand LIC bits and how to maximize ROP.
- **Coring Fluid Specialist.** This person's key responsibilities include directing pilot testing (i.e., mud-composition design), proper mud mixing, transportation to rig if needed, pit and equipment cleaning, displacement to coring fluid, mud maintenance, and sampling.
- **(Two to Four) Core-Analysis Laboratory Field Technicians.** These individuals are responsible for core mark-up and cutting; QC sampling and preservation; and nitrogen purging, capping, and shipping. If the core is to be slabbed and plugged in the field, more people are required.
- **Operator Core-Analysis Specialist.** This person must be at the core-analysis laboratory when the core arrives to ensure that appropriate sampling is consistent with project objectives, that proper laboratory techniques are used, and that the chance of water contamination and/or loss of connate water is minimized.

Chapter 3

Low-Invasion Coring Bit Selection

3.1 General Features

The early research in the area of low-invasion coring (LIC) focused on bit design. Several SPE technical papers present studies of the mechanics of cutting core and the changes to the coring systems that possibly reduce mud-filtrate invasion (Tibbitts and Radford 1985; Rathmell et al. 1995). In these studies, increasing the ROP is considered the most important goal, but rerouting the mud-fluid flow from throat discharge to face discharge to minimize mud-filtrate invasion is also important (**Figs. 3.1 and 3.2**). Some of the key features of the new bits are as follows:

- As a part of maximizing ROP and cutting efficiency, PDC cutters are used. The bits are very lightly set (i.e., minimal cutters) and are designed with the cutters on blades to create large junk slots that allow cuttings to be efficiently flushed from the bit. The tradeoff to some of these features is a shorter bit life.
- The bit profile is changed from round to parabolic or semiparabolic.
- Bits are designed to minimize downhole torsional and lateral vibrations. *Bit whirl* refers to a bit with imperfect rotation that will deviate from the center drilling axis. Lateral vibration may cause a hole slightly larger than the bit diameter. In coring, such imperfect rotation leads to undergauging a core, which may then fall out of the catcher and reduce core recovery. Lateral vibration also leads to more core breakage. With a stability-optimized bit, the energy not wasted in lateral movements is available to drilling efficiencies such as improved ROP.
- The catcher shoe is extended into the mouth of the core bit so that the core is protected—as quickly as possible after being cut—from exposure to high-velocity mud.
- The new low-invasion bits are designed with face-discharge ports to direct the majority of the mud out of the face of the bit and away from the core. *Note: Legacy designs were all throat discharge bits, meaning the mud exited the coring assembly directly at the point where new core was being cut, which increased mud-filtrate invasion.* The newest version of these low-invasion bits may have 85% of the mud flow directed out of the face-discharge ports, leaving only sufficient mud flowing from the throat to ensure cuttings removal and cooling.
- Throat gauge cutters are eliminated to avoid removing freshly deposited filter cake, which increases filtrate invasion.

The bit design changes were very effective. Coring ROP increased significantly. However, consistent low invasion is not achieved until the new bits are coupled with specially designed LIC fluids (see Chapter 5).

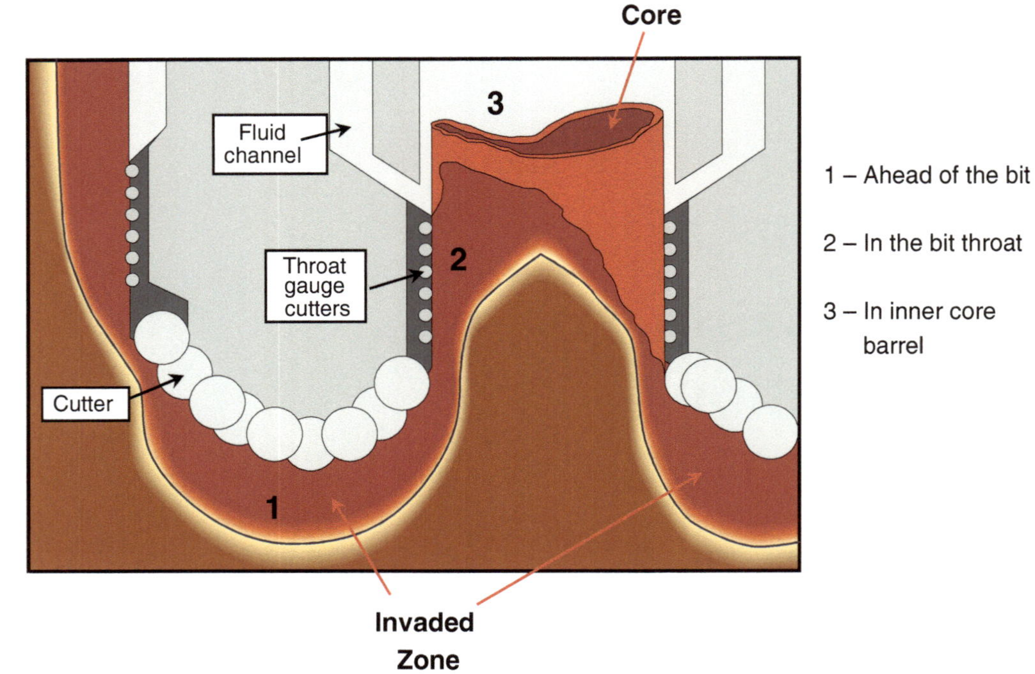

Fig. 3.1—Filtrate invasion pattern on standard core bit. Courtesy of Baker Hughes.

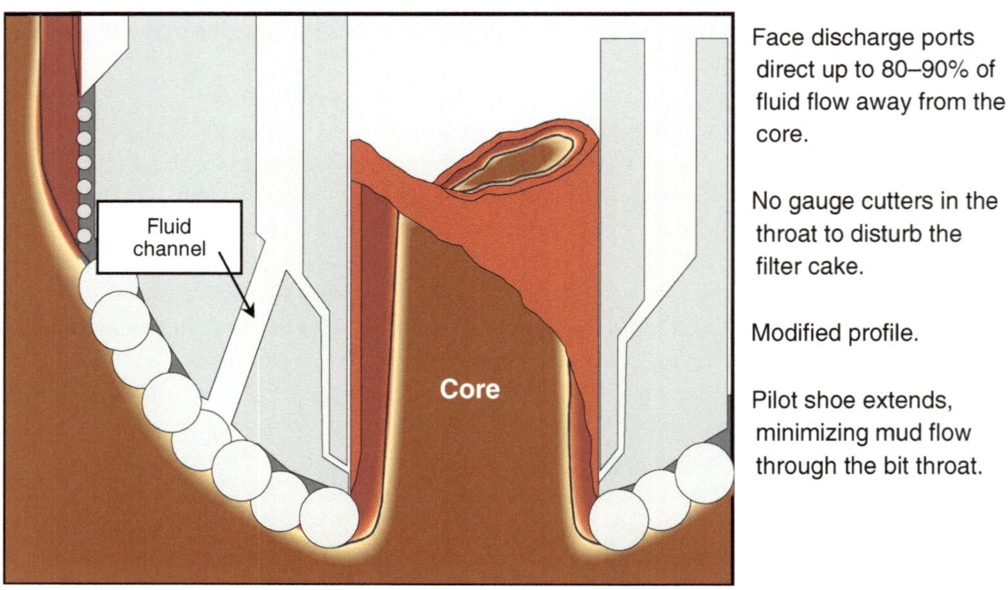

Fig. 3.2—LIC bit modified to reduce filtrate invasion. Courtesy of Baker Hughes.

Note: Most of the early low-invasion PDC coring bits were developed as Baker Hughes Inteq (BHI) bits and equipment. Much of the discussion in this monograph is based on those coring bits. Halliburton also developed a series of low-invasion-style bits based on the CD-93 bit discussed in Clydesdale et al. (1994).

3.2 Bit Selection vs. Rock Type

3.2.1 BHI Low-Invasion Bits. Low-invasion bits from BHI will have nomenclature BHC-XXX, yet can only be determined by the hydraulic configuration and bit options associated with each design. Legacy low invasion bits from BHI had the nomenclature AXC-XXX. The "AXC" stood for an anti-whirl design. All "AXC" series bits have been replaced with the "BHC" series based on Hughes Christensen's Genesis™ drillbit platform. Bit technology improves continually; therefore, one should work with the bit company on the details of the newest low-invasion bits. *Note: Although LIC bit sizes vary greatly, smaller-diameter cores have a smaller required cross-sectional area to be able to realistically achieve low mud-filtrate invasion.* The large-diameter bits cut more noncore rock than their smaller counterparts and, therefore, are usually unable to achieve the high ROP necessary for low mud-filtrate invasion. The exception is poorly consolidated rock that is cut very easily and in which larger-diameter cores are preferable.

BHC-407. The BHC-407 (with seven blades) is an improved version of the legacy low-invasion AXC-413 design, incorporating the newest PDC cutters and with added "BRUTE" (Backups that are Radially Unaggressive and Tangentially Efficient) cutters behind the main row of cutters on each blade. These extra cutters control the depth-of-cut with each rotation and make it even easier to see WOB even with modest ROP. The AXC-413 was specifically designed for use in poorly consolidated sandstone (0- to 5,000-psi unconfined compressive strength). Flow through the throat is limited to only 10-15%. Because of the restricted mud flow through the bit throat, the minimum acceptable flow rates required for bit cleaning usually are approximately 100 gal/min. For extremely loose sand, even this flow rate may be too high. In some areas, 60 gal/min is needed to reduce core loss.

Halliburton (Security/DBS) Bits. Security/DBS bits are somewhat similar in design to the BHI BHC-series because they primarily use 0.5-in. PDC cutters lightly set on blades, eliminate extra gauge cutters in the throat, and incorporate antiwhirl features through their bit-stabilization system. The catcher shoe is not extended, but its end fits into a slot in the throat of the coring bit. This insertion creates a very narrow, convoluted path for mud flow and results in directing almost all the mud out of the face-discharge ports. To overcome the problem of cleaning cuttings out from under the bit, the face-discharge ports are reshaped and angled sharply outward to create a Venturi effect designed to vacuum cuttings out of the way. The blades are angled forward in a spiral fashion, and the profiles are round/flat. The FC264LI is a six-bladed bit, somewhat similar to the AXC-412M. The FC284LI is an eight-bladed bit similar to the AXC-425, but with a flatter profile.

3.2.2 Bit Selection Preplanning. The usual approach is to use all available data from nearby wells (i.e., drilling records, previous coring rates, and core mechanical properties) to make the best bit selection. Because these data are frequently not available, it usually is advisable to bracket the primary choice by having at least one harder-rock bit and one softer-rock bit available at the rigsite. The coring project manager must work with the bit vendor, coring engineer, and drilling engineer to maximize the likelihood of having an optimal-ROP core bit on site. This group should also estimate the number of core bits needed to complete the coring operations. Low-invasion bits are generally built for high ROP but at the expense of bit life. One should not plan on getting more than two or three 90-ft coring runs from each low-invasion bit. The worn bits are probably suitable for additional regular coring, but dull enough not to give optimum ROP for LIC.

The lead times required for manufacturing and shipping bits to the rigsite can vary from 10 days to 6 weeks. One must inquire about core bit availability early, and accordingly plan ahead. The

Fig. 3.3—BHC406 LIC bit. Image courtesy of Baker Hughes.

threads and shank design must be matched to the outer core barrel (with a high-torque barrel preferred). Also, if working in poorly consolidated sandstone, the shank must be matched to the full-closure core catcher. If changes are needed, the procedure takes approximately 2 days in a core-bit manufacturing shop.

Remember, low ROP probably means mud-filtrate invasion to the core center, but valid connate water saturations can still be obtained using 100% OBM, and in-situ wettabilities can be obtained using bland WBM.

Chapter 4

Other Downhole Hardware

4.1 Inner Core Barrels

Inner core barrels made of aluminum are generally preferred for low-invasion coring (LIC). They are stronger than fiberglass (which experiences occasional failures at connectors) and can tolerate higher temperatures. Aluminum inner core barrels are also more rigid and less prone to allowing core breakage caused by flexing. Aluminum inner core barrels can be readily cut into sections with four-wheel pipe cutters. Saw blades that cut both aluminum and rock are difficult to find. Blades of bonded grit work, but heat up the core as far as 9 in. from the cut location. Fiberglass inner core barrels make it easier to cut the core into sections by using a saw, but there is a concern about local heating altering fluid saturations at or near the cut points. Both types of inner core barrels are transparent to gamma rays; therefore, the core sections can be logged without depreserving the core.

> Aluminum inner core barrels are not compatible with high-pH WBMs at high temperatures. After noting extreme corrosion while coring a deep, hot well in the South China Sea, a small set of laboratory data was generated (unpublished). Samples of aluminum inner core barrel were hot-rolled at 325°F for 24 hours in a KCl/Polydrill™/Rheostar™ mud with an initial pH of 8.0, 9.0, and 10.0. The pH 8.0 and 9.0 samples showed little weight loss, but the pH 10 sample lost more than one-third of its mass. This effect is worse while coring, because the high-pH fluid in contact with the inner core barrel is continually replaced while circulating the coring fluid. These data only cover one set of conditions, but establish a warning to do preliminary laboratory work if WBM is to be used at high temperatures. Note, also, that solubilized aluminum is very likely to cause core or formation damage. The high pH of the internal brine phase of an OBM only slowly damages aluminum inner core barrels because oil is the continuous phase, and the aluminum is generally covered with an oil and surfactant film.

When inner-barrel connectors are needed (i.e., cores greater than 30 ft), the connectors should have a built-in rotating sleeve stabilizer to minimize inner core barrel rotation in case of contact between the inner and outer core barrels (especially in high-angle holes). It is also preferred that these connectors have removable windows to facilitate breaking the core into 30-ft sections on the rig floor (see Chapter 6 for more information).

4.2 High-Torque, Outer Core Barrels

Because one goal of an LIC is to maximize ROP, high-torque, outer core barrels like the BHI HT-30 and HT-60 series are preferred. These barrels are designed with high-strength, double-shouldered connections for better performance under high-torque conditions like high-angle wells, downhole motor coring, long-core barrels, and high-rev/min conditions.

4.3 Mud Motors

In harder rock, where high ROPs are difficult to obtain, using a downhole positive-displacement mud motor to increase the bit rev/min can extend the range of formations successfully cored with low mud-filtrate invasion. Doubling the rotary speed of a core bit from 100 to 200 rev/min roughly doubles the ROP. Antiwhirl low-invasion core bits are generally stable at 200 rev/min. Rathmell et al. (1998) discusses more details of a field application of this approach.

Because the ball that diverts the mud flow from flushing the inner core barrel to passing between the inner and outer barrels cannot pass through the motor, the ball must be premounted below the motor with some style of drop-ball sub or tool. The ball is released from both the drop-ball sub and a downhole-activated flow diverter (DAFD) tool by increasing or surging the flow once the hole has been circulated clean. To be able to circulate bottoms up after coring without rotating the core, a ball-activated circulating sub must be run above the motor.

The motor should be capable of high torque, such as BHI's Navi-Drill Mach 1 (or the best current motor). If appropriate WOB cannot be reliably generated, and because of hole geometry, a thruster can be run above the mud motor.

4.4 Core-Catcher Assemblies

LIC bits use extended shoe assemblies to minimize the amount of mud flow in the throat area and thereby protect the core within an inch after being cut. For consolidated core, a standard spring-type catcher is used. The core can easily push through the catcher into the inner core barrel; but if the core moves back toward the bit, the catcher slips down, is compressed onto the core by the lower shoe, and grips the core tightly with its surface of tungsten carbide grit.

The spacing between the core catcher shoe and core bit should be carefully determined based on the estimated bottomhole circulating temperature. The spacing is set on the surface to compensate for differential expansion between the steel outer barrel and the aluminum inner core barrel. The coring engineer will have a chart to determine the proper spacing for a given core barrel length and bottomhole circulating temperature. For high-angle holes and long core barrels, high-strength outer core barrels should be used.

For poorly consolidated cores, a full-closure type catcher is needed. Unconsolidated rock is too weak to be held inside the inner core barrel by a standard split-ring, spring-type catcher. The best system currently available is the BHI Hydro-Lift™ system. When coring is complete, a second ball is dropped that diverts mud flow to lift the entire inner core barrel slightly for a pair of powerfully spring-loaded, clamshell-type catchers to be slammed together to seal the core in the barrel. Halliburton (Security/DBS) has a full-closure core-catcher system that employs additional flow rate (i.e., pressure) to crush a device that looks like an aluminum soda can to seal the bottom of the assembly. It is a creative concept, but based on some experiences, it tends to activate early if the pressure spikes; and it does not usually completely seal if there is any rock in the throat of the "can."

Full-closure core-catcher systems should be backed-up by a secondary standard split-ring catcher. If the coring run ends in a few feet of harder rock, the standard catcher ensures full recovery. In fact, when cutting sand and shale sequences, it is a reasonable procedure to attempt terminating the coring with a couple of feet of shale across the catcher. This approach may not hold true in extremely weak sands. The extra vibration generated while cutting shale is thought by some to damage the soft sands in the core barrel.

Damage to the core can be identified by way of CT scans while the core is still preserved in the 3-ft sections of the inner core barrel. With frozen core, these CT scans also allow the bedding planes to be defined for slabbing to show maximum bedding dip on the slabbed surfaces. In soft sediments, the CT scans may be the only way to view gas-expansion damage because the slabbed sections typically do not reveal this type of damage.

Chapter 5

Low-Invasion Coring Fluids

5.1 Overview

Early low-invasion coring (LIC) projects showed that redesigning the bit was only part of the answer to consistently achieving uninvaded core centers; therefore, research on improving and fully understanding coring fluids was also needed. The work was performed as Joint Industry Projects (JIPs) under the Drilling Engineering Associations DEA 23 and DEA 57 and is partially summarized in Walker and Black (1993). These tests were performed (1990–92) at TerraTek's Drilling Research Laboratory in Salt Lake City, Utah. The goal was to reduce the invasion of mud filtrate into a core by forming a very low-permeability filter cake as quickly as possible. It was concluded that 50 to 75% of the improvement needed to achieve low mud-filtrate invasion depended on having a premium coring fluid. The research focused on spurt loss—the first few seconds after rock is cut when mud rushes into the pores, and the solids bridge off and form a filter cake.

Three different sandstones were tested as follows:

- Gold sandstone 50 md
- Berea sandstone 600 md
- Bentheimer sandstone 2,500 md

The 1-in. diameter core plugs were sealed on the circumference with epoxy, mounted in the drilling apparatus, and then saturated with brine. The plugs were then drilled with a 1-in. two-cutter PDC microbit, which consumed the plugs completely. The drilling was done at realistic temperatures and mud overbalance pressures. As the bit exposed new rock and a new filter cake began to form, the spurt loss of the mud displaced the mobile brine in the core. By accurately measuring the displaced brine, the spurt loss or filter-cake-forming efficiency of the mud was measured. A wide range of mud-composition variables was tested.

Several key coring fluid rules of thumb were developed (Walker and Black 1993):

- OBM produces much lower spurt losses and much-lower-permeability filter cakes than WBM.
- A premium filter cake requires clay as a component (bentonite in WBM or amine-treated bentonite in OBM).
- High concentrations of soft, deformable, colloidal fluid-loss materials significantly reduce spurt loss (fluid-loss polymers in WBM, and asphaltic materials or oil-soluble polymers in OBM).
- High-solids loadings (>10% by volume) are preferred for rapid filter-cake formation.
- Fine (3 to 10 μm median size) bridging materials produce minimal spurt loss for a wide range of rock permeabilities. Sized calcium carbonate is readily available and very well suited as an effective bridging material.

Subsequent research and field experience resulted in the following additional findings:

- Displacing to a clean, site-specific coring fluid increases the probability of success better than trying to adapt existing field muds.
- Biocides are critical in WBM. Bacteria can destroy filter-cake quality in a variety of ways. (See subsequent discussion in Section 5.2.)
- It is best to fill the mud pits with sufficient mud to complete all coring operations without having to make dilutions or additions.
- Tracers in the mud filtrate are needed to determine the degree of mud-filtrate invasion that occurred in the core. A uniform tracer concentration must be maintained throughout the mud system and throughout the time period when it is being used.
- The coring mud should be designed to result in sufficient core from which substantial amounts of high-quality data can be obtained, even if low invasion is not achieved. For example, if WBM filtrates contain no surface-active materials, then even filtrate-invaded cores can give reasonably accurate wettability values. Also, if an OBM contains minimal concentrations of weak surfactants, then even a mud-filtrate invaded core can yield accurate water saturations.

5.2 Water-Based Coring Fluids

An LIC fluid formulation must be adapted to the conditions of each individual well or reservoir, but a reasonable standard formulation is shown as follows:

Water	as needed
Biocide	0.5 ppb (if using 25% glutaraldehyde)
Bentonite	7 to 8 ppb
Starch	7 to 8 ppb (potato starch preferred)
Low-viscosity polyanionic cellulose (PAC) polymers	1.75 to 2.0 ppb
Xanthan gum	as needed (approximately 0.4 ppb)
KCl	6% or more, as needed
Calcium carbonate	50 to 100 ppb, marble, 3- to 10-μm median size
Barite	as needed
Tracer	850 ppm deuterium oxide (D_2O) or 1.0 ppb KBr or NaBr or 0.2 ppb KI or NaI or 220 millicuries of tritiated water per 1,000 bbl (Radioactive tracers involve many regulations)
KOH	if needed, pH 7.5 to 8.5

The mud components are listed in the order in which they must be added to form the whole mud. It is strongly advised that pilot testing with locally available products be used to refine the formulation. The tracer is usually added only after a successful displacement of the drilling mud by the coring fluid is achieved, especially if the tracer is either difficult to acquire in the field or is as expensive as D_2O. Because of high-polymer concentrations and high-solids loading, the final rheology (i.e., viscosity as a function of shear rate) of water-based coring fluids can be fairly high relative to standard drilling fluids. Because typical coring-fluid flow rates are less than 300 gal/min, the extra viscosity should not cause unacceptable pumping pressures.

Typical specifications for water-based coring fluids are as follows:

- Target pH: 7.5 to 8.5.
- Target high-pressure/high-temperature (HP/HT) fluid loss (30 minutes at 500 psi differential pressure, 200°F): < 8 cm^3 (6 to 7 cm^3 is frequently possible in the field). When the mud

components for low mud-filtrate invasion are present, HP/HT fluid loss correlates closely to spurt loss. Lower values are always better, if readily achievable.
- Soluble calcium: 100 to 300 ppm.
- No surface-active materials should be added, such as lignosulfonates, lignites, drilling detergents, sulfonated asphalt, or lubricants.
- The rheology of the fluid should be adequate for hole cleaning and weight-material suspension.

Important aspects of the individual components are as follows:

- **Biocide.** Starch and xanthan gum polymers are very susceptible to degradation by bacteria, especially in low-pH fluids. PAC polymers are moderately susceptible to bacterial attack. Even though mud encounters a harsh range of pressures and temperatures, bacterial populations can increase by a factor of 10 during each 24-hour period. Even modest bacteria problems (i.e., no smell or alkalinity buildup) can cause significant fluid-loss-control and rheology problems. Even if bacteria are discovered and killed, they can leave behind catalytic enzymes that continue to destroy polymer for days. The best approach is to pretreat all make-up water with biocide and add a daily maintenance dose. A good choice is 25% glutaraldehyde, which is available from all of the major drilling-fluid companies. Glutaraldehyde kills a wide range of bacteria, is stable if cement is drilled, and can be monitored with a field test kit. If test kits are used, the target concentration in the mud filtrate is 300 ppm. Otherwise, make the initial dose 0.5 ppb, and add a daily maintenance dose of 0.05 to 0.1 ppb during coring. Note: Other biocides may be used if glutaraldehyde is not available. However, the dose rate should be approximately double the manufacturer's suggestion to allow for consumption, surface adsorption on cuttings, and other routes of depletion.
- **Bentonite.** Bentonite is used for fluid-loss control, viscosity, and filter-cake quality. Bentonite is critical to achieve low spurt-loss values. Muds made only with polymer and calcium carbonate have moderate-to-high spurt-loss values. The bentonite should be prehydrated in fresh water, without using caustic. Bentonite does not fully disperse in a mud pit that still contains some residual KCl fluid. The target concentration is typically 7 to 8 ppb. If the quality of the bentonite is good, then concentrations with higher than 8 ppb quickly result in an undesired increase in viscosity. Concentrations lower than 6 ppb are too low to give a good base filter cake. "Bentonite equivalent" from drilled solids is not as good as commercial bentonite.
- **Starch.** Starch by itself produces a thick, soft filter cake, but a 4:1 combination with low-viscosity PAC produces high-quality filter cakes, low HP/HT values, and low spurt-loss values. Potato starch is preferred, but corn and tapioca starches have been used after appropriate pilot testing. *Note: Starch has an upper-temperature limit of 220 to 225°F, and requires an aggressive biocide program.* Above 225°F, low-viscosity PAC can be used alone (approximately 2.50 to 2.75 ppb). Above 275°F, specialty polymers or OBM must be used.
- **PAC.** PAC polymers are used for fluid-loss control and filter-cake quality. Regular-viscosity PACs give good fluid-loss control, but too high a viscosity. Even low-viscosity PAC generates noticeable viscosity increases when used at high concentrations. New ultralow-viscosity PAC can alleviate these problems, but approximately twice as much polymer may be required. The temperature limit for PAC is approximately 275°F.
- **Xanthan Gum.** Xanthan gums produce better rheology for solids suspension and hole cleaning than PAC. In most cases, a treatment of 0.25 to 0.70 ppb is adequate. The temperature limit for xanthan gums is approximately 275°F.
- **KCl.** Potassium chloride is used for core preservation, formation-damage control, and borehole stability. KCl inhibits the swelling and dispersion of some water-sensitive clays. The minimum recommended concentration is equal to the formation total salinity, but higher levels are acceptable and are frequently needed for borehole stability. A mud filtrate lower in salinity than the formation brine can either cause clay swelling or give a chemical/environment change

called "salinity shock," which can mobilize migratable clays. Borehole stability is also important. High-core recoveries depend on the hole being within gauge as much as possible because the hole acts as a journal bearing to stabilize the coring assembly. Assuming enough total salt is present (a minimum of 6% KCl is generally preferred), then hole stability is significantly improved when the ratio of potassium to sodium in the mud filtrate is high (at least 2:1 is needed, 3:1 is better). Tracking the potassium concentration is best determined by the American Petroleum Institute (API) ion-selective-electrode method. Sodium is difficult to track at the rig, but a good estimate of the potassium-to-sodium ratio can be made by reviewing the potassium-to-total-chloride ratio (K/Cl). If the K/Cl ratio (ppm/ppm) is 0.70, then the potassium-to-sodium ratio (K/Na) is approximately 3:1; and if the K/Cl ratio is 0.596, then the K/Na ratio is approximately 2:1.

Note: Displacing an uncased hole from a freshwater fluid to a KCl fluid can result in borehole-stability problems if water-sensitive shales are present. If a KCl coring fluid is to be used, then the hole section above the core point should be drilled with a KCl mud.

- **Calcium Carbonate.** Calcium carbonate is used as pore-throat-bridging particles to help quickly establish a thin, tough, low-permeability filter cake that minimizes the invasion of mud filtrate into the core and the formation. The critical characteristic of this product is the median particle size, which should be in the 3- to 10-µm range. In addition to particle size, the quantity is very important. It is preferred to have at least 50 to 60 ppb of the bridging material, and up to 100 ppb is helpful. Whenever possible, the total solids content in the mud should be at least 10% by volume. At these levels, sandstones of up to 3-darcy permeability have been successfully cored and have shown low-mud-filtrate invasion. Very fine calcium carbonate (325 to 800 mesh) is widely available around the world because it is used as a filler when manufacturing paper. Calcium carbonate ground from marble is preferred because it is harder than limestone and retains its size distribution better when circulated through the mud pumps and the bit.
- **Barite.** Barite is finely ground barium sulfate and is used as a weighting agent. Its high specific gravity (4.2 g/cm^3) allows mud density to be raised with a minimum addition of solids. Calcium carbonate can be used to raise the density of coring fluids to a maximum of approximately 10 lbm/gal. Barite should be used in combination with calcium carbonate for higher densities. The average particle size for most barites is 20 to 25 µm.
- **Tracers.** These are covered in Section 5.4 of this chapter.

5.2.1 Adapting a Water-Based Field Mud for Coring. It is strongly recommended that water-based LIC be done by first displacing to a site-specific premium coring fluid as previously described. Field muds are very likely to cause difficulties due to surface-active components, pH, spurt loss, filter-cake quality, and proper amount and size of bridging solids. However, if costs, logistics, or some other consideration precludes the use of a site-specific coring fluid, there are steps that can improve the likelihood of success with a field mud. Remember, there is no second chance if mud filtrate invades to the center of the core. Wettability may be altered if surface-active materials are present, which means that ROP becomes very critical. If coring rates of at least 30 ft/hr (preferably higher) cannot be achieved, displacing to an optimized coring fluid is the preferred option.

Alternative. An intermediate option is to estimate the amount of premium fluid needed for coring, and make up only that amount and circulate it across the bit only while coring. This option is sometimes used in deepwater coring projects when displacing the entire circulating system (approximately 5,000 bbl) would be prohibitively expensive; 1,000 bbl of coring fluid is a common amount but is very dependent on the amount of coring planned and the ROP achieved. In an 8.5-in. hole, circulation rates of approximately 300 gal/min are standard for consolidated rock. For poorly consolidated rock, the circulating rate probably does not exceed 100 gal/min. Mixing the used coring fluid into the main drilling fluid is usually not a problem.

The goals outlined for designing a premium coring fluid also apply when adapting field muds for coring.

- A few days before coring, stop adding caustic and allow the pH to decrease to the preferred range of 7.5 to 8.5. The higher the initial pH, the longer this process takes. *Note: This approach does not work for lime or silicate muds.*
- An HP/HT fluid loss of < 8 cm^3 is still a good target value, but is harder to achieve with a field mud full of drilled solids without the viscosity increasing substantially. A whole-mud-system dilution of 10 to 15% can reduce viscosity enough to allow the addition of adequate amounts of starch and PAC (4:1).
- A field mud probably contains surface-active agents like thinners, dispersants, detergents, lubricants, and sulfonated asphalt. Where possible, stop additions of these items a few days before coring, which allows time to, at a minimum, partially deplete these components in the mud system. As previously discussed, a dilution of the entire mud system is preferable.
- Adding KCl to the mud may be considered under certain circumstances. It helps with borehole stability and core preservation. However, there are certain negative considerations. Adding salt to a freshwater mud can cause a large increase in viscosity. If there is substantial footage of open hole present, then changing from freshwater to KCl can cause sudden and severe hole-stability problems in shale intervals. In all cases, pilot testing the conversion to a KCl mud is recommended. A good first step is to dilute the mud to a bentonite content of < 12 ppb.
- Because field muds are already loaded with solids, it is difficult to add adequate amounts of fine calcium carbonate without causing mud-weight and viscosity problems. Each 7 ppb of calcium carbonate raises the density of the mud by approximately 0.1 lbm/gal. Adding 50 to 60 lbm/bbl of calcium carbonate into the mud is very difficult, but at least 30 lbm/bbl should be attempted. Several options may be considered. The density of the surface system can be centrifuged back either 0.5 lbm/gal or more, and the weight replaced with fine calcium carbonate. If a weight-up is scheduled shortly before coring, the density of the mud can be increased with calcium carbonate. If the mud pits are low, new mud weighted exclusively with calcium carbonate can be added. As previously discussed, a 15% dilution can help with several mud properties and can allow for adding calcium carbonate. Different rigs and different projects have different constraints. In summary, a number of options are possible to achieve the desired objective—increasing the fine calcium carbonate content of the mud system.

5.2.2 Determining the Amount of Calcium Carbonate in Mud. There may be times when it is desired to determine the amount of calcium carbonate in WBM. One example from a past project involves the scenario of two core points 1,000 ft apart. The coring fluid may be optimal for the first core, but after drilling the next 1,000 ft, half of the original calcium carbonate may be depleted. Appendix D contains a method for determining the amount of calcium carbonate in whole WBM. It is simple, accurate, and uses only chemicals already available at the rig. The procedure involves using hydrochloric acid to dissolve the calcium carbonate, and then titrating for the soluble calcium. Because OBMs may contain high concentrations of calcium chloride, this method has not yet been adapted to determine the amount of calcium carbonate in an OBM. It may be possible to break the OBM emulsion as in a normal API test, and filter out the solids (including the calcium carbonate) before running the test described in Appendix D.

5.3 Oil-Based Coring Fluids

OBMs are inherently better for LIC, especially when water saturation is the main objective. They have a lower spurt loss, produce a lower-permeability filter cake, and usually result in higher ROPs. Even if some filtrate invasion occurs, accurate water saturations can be obtained in connate water zones (if the filtrate/brine IFT is high—the formulation given in the following subsection gives an

IFT of approximately 17–18 dynes/cm). However, if wettability is the main objective, then OBMs should only be used if the ROP is high enough to ensure no filtrate invasion to the center of the core (>15 to 30 ft/hr). Field muds can be modified for coring, but it is strongly preferred to displace to a site-specific, properly designed LIC fluid like the one described in Section 5.3.1.

5.3.1 Purpose-Built, Oil-Based Coring-Fluid Parameters.

- The mud should have a very low spurt loss (filtrate front advancing slower than the ROP).
- The mud should have an HP/HT fluid loss of 2 to 4 cm^3, or less. The lower the HP/HT number, the better.
- The mud should contain 60 to 100 ppb of calcium carbonate with a median size of 3 to 10 µm.
- The mud should have a very low water content. Less than 2% by volume is preferred.
- The only surfactants the mud should contain should be carboxylic acid emulsifiers in low concentrations (2 to 3 ppb).
- The IFT between the filtrate and the formation brine (or a synthetic laboratory brine) should be >7 dyne/cm, with a goal of >10 dyne/cm. Any commercial laboratory with a spinning-drop interfacial tensiometer can determine the IFT.
- The excess lime in the mud should be kept at approximately 1.0 ppb.
- The electrical stability (ES) of the mud (< 2% water) should be maintained at 2,000+ V.
- The mud must contain an oil-soluble tracer to track the depth of mud-filtrate invasion into the core.

The mud formulation is adjusted to fit the conditions for a specific well, but the following design parameter ranges have been successful for a number of projects.

Base oil	as needed
Organophilic clay	6.0 ppb
Lime	2.0 to 3.0 ppb
Emulsifier-dimer/trimer fatty acid	2.0 ppb
Blown asphalt, *or*	25 to 35 ppb
Oil-soluble polymer	varies, as needed
Gilsonite	up to 5 ppb
Calcium carbonate	60 to 100 ppb (3- to 10-µm median size)
Barite	as needed
Tracer (1-bromonaphthalene)	100 ppm by weight of liquid phase

If blown (i.e., air-oxidized) asphalt is not available, then a combination of oil-soluble, fluid-loss polymers and up to 8 ppb of Gilsonite can be used. Gilsonite concentrations higher than 8 ppb can lead to unacceptably high gel strengths. In the Gulf of Mexico area, asphaltic products can cause problems with environmental testing, and all fluid-loss control may need to be done with oil-soluble polymers.

Other considerations for oil-based coring fluids include

- Be sure to use a premium wet-processed, high-purity clay, and pilot-test its initial viscosity yield and its reaction to minor water contamination. Standard organophilic clays are not fully coated with amine and can react with even trace amounts of water to double the expected viscosity.
- The HP/HT fluid loss target for an all-oil coring fluid is 2 to 4 cm^3, or less; 1.5 to 2.0 cm^3 is frequently achieved in the field. As with WBM, when the components for a low spurt loss are present in the mud, the mud-filtrate invasion character of the mud relates directly to the HP/HT fluid loss.

- Water content. The mud should have very low water content. Remember that, often, the primary goal of LIC is to obtain accurate in-situ water saturations. Water in the mud, if water tracers are not used, can somewhat increase the uncertainty as to whether the water extracted from the core is connate or introduced. The filter cake excludes tightly emulsified water once it forms, but a small amount of water may be carried into the core with the first spurt of whole mud. The preferred water content is < 2% by volume. If there is a concern about confusing water from the OBM with formation brine, then one of the water-based tracers described in Section 5.4 of this chapter can be added.

5.3.2 Controlling Water Contamination in OBM. Key procedures for keeping water contamination to a minimum are as follows:

1. The mud must be mixed in clean, dry tanks.
2. If mixed off-site, the mud must be transferred to the rig in clean, dry vehicles. The vehicles (i.e., either tank trucks or workboats) must be inspected before loading the coring fluid to ensure they are dry.
3. The mud pits, piping, and solids-control equipment at the rig must be cleaned (i.e., pipes and pumps must be flushed with base oil).
4. The water lines must be shut off to the rig floor, mixing equipment, and solids-control area.
5. The mud system must be protected from rain contamination.
6. The previous mud in the hole must be carefully and thoroughly displaced.

We offer the following comments on oil-based coring fluid components:

- **Emulsifier.** Any water in the coring fluid must be tightly emulsified to prevent it from moving through the filter cake and altering connate water saturations. The preferred emulsifiers are carboxylic acid materials (also called fatty acids or tall oil). They are effective water emulsifiers, yet they have reasonably high IFTs that prevent them from being very effective surfactants (Dawe and Murdock 1990; Reed et al. 1994). Being ineffective surfactants means that if mud filtrate with these emulsifiers invades the core, these surfactants are incapable of mobilizing connate water and altering the connate water saturations determined from the Dean-Stark water extractions. The best materials are the dimer/trimer fatty acids used as low-shear-rate rheology modifiers. These materials, if present at no more than 2 to 3 ppb, have very little ability to mobilize and displace the connate water in a core. No amine-based emulsifiers should be used. Standard field OBMs often have excess concentrations of very aggressive surfactants and are not preferred for coring when accurate water saturations are a major goal of the coring project.
- **IFT.** As previously mentioned, the IFT between the coring fluid filtrate and the formation brine should be kept as high as possible to prevent connate brine from being mobilized and displaced from the core (Dawe and Murdock 1990; Reed et al. 1994). If the IFT is kept high (preferably > 10 dyne/cm), connate water is unlikely to be mobilized, and accurate water saturations can be determined even if the core center is invaded with mud filtrate. The dimer/trimer fatty acids listed in Section 5.3.1 typically produce IFTs in the 15 to 18-dyne/cm range. See Appendix I for additional discussion of the importance of IFT and the other parameters that affect connate water saturation and wettability.
- **Lime.** Lime is needed to keep the emulsifiers in the calcium-soap form, which is a more efficient type of emulsifier and gives higher IFT. An excess of approximately 1.0 ppb should be maintained. *Note: Lime is difficult to measure accurately in OBM because the surfactants tightly coat any undissolved material.*

- **ES.** ES is a measure of the quality of OBM emulsions. With a very tight emulsion, no water from the whole mud can pass through the filter cake and contaminate the core. To ensure that any water in an "all-oil" coring fluid is thoroughly emulsified, the ES goal is > 2,000 V. A digital chip-calibrated Fann 23D-type instrument should be used, not the old-style analog models.

5.3.3 Adapting Field OBMs for Coring. Typical field OBMs are generally good coring fluids (i.e., lower mud-filtrate invasion, better filter cake, and higher ROP), but may be difficult to adapt for LIC if water saturations are the primary goal. Most field OBMs carry very high loadings of strong surfactants, which may mobilize residual water if significant mud-filtrate invasion occurs. Also, field OBM usually contains significant volumes of emulsified water that may enter the core during the mud spurt and increase the uncertainty of accurate connate water saturations. Typical oil/water ratios for field OBM are in the range of 70:30 to 85:15. There are a few mud treatments that improve the probability of obtaining low mud-invasion cores, but if the ROP is not at least 15 to 30 ft/hr, then consider displacing the field OBM with a premium low-water-content, oil-based coring fluid.

Less Surfactant. Several days before coring, the wetting agent/emulsifier additions to the mud system should be minimized and changed to fatty-acid-type products. If feasible, measuring the IFT (performed either by a commercial laboratory or university that has a spinning-drop interfacial tensiometer) between the mud filtrate and simulated formation brine determines how likely the mud filtrate is to mobilize water in the core. The goal is an IFT >10 dyne/cm. *Warning: The mud must contain enough surfactant to keep the barite, calcium carbonate, and drilled solids in an oil-wet condition. If too little wetting agent/emulsifier is present, then the viscosity of the mud can significantly increase because of an unstable emulsion or water-wet solids.* Pilot-testing additions of barite to a mud sample until the mud suddenly becomes dramatically thicker can roughly estimate the excess surfactant in the mud. The mud should be able to tolerate at least 10 ppb of barite addition with only minor increases in rheological parameters. If further drilling is planned after coring, be sure a strong emulsifier is added to the mud before drilling ahead.

Tracer in Water Phase. Because a high-water-content OBM may introduce water into the core during the spurt loss, it is prudent to add a tracer into the water phase of the mud. D_2O, bromide ions, or iodide ions work well as outlined in Section 5.4. The quantity of tracer should be determined based on the barrels of water phase in the mud system. The salts should be predissolved in a small amount of water and added in uniformly over one full circulation of the mud system.

HP/HT. The HP/HT fluid loss target of 2 to 4 cm^3 may be achievable even in a field OBM. A 5 to 10% dilution with base oil may be sufficient in a solids-loaded system to allow the addition of adequate amounts of blown asphalt and Gilsonite, or oil-soluble polymer. The higher water content of field muds makes it easier to achieve a low HP/HT fluid loss. Pilot testing is recommended.

Calcium Carbonate. As with WBM, it is desirable to get at least 30 ppb of fine calcium carbonate (preferably 50 to 60 ppb) into the mud system to reduce spurt loss. Because field OBMs are already loaded with solids, it is difficult to add adequate amounts of fine calcium carbonate without causing mud-weight and viscosity problems. Each 7 ppb of calcium carbonate raises the density of the mud by approximately 0.1 lbm/gal. Several options may be considered. While running casing, waiting on cement, or any other delay, the density of the mud in the surface system can be centrifuged back 0.5 lbm/gal or more, and the weight replaced with fine calcium carbonate. If a weight-up is scheduled shortly before coring, the mud density can be increased with calcium carbonate. If the mud pits are low, then new mud weighted exclusively with calcium carbonate can be added. As previously discussed, a 5 to 10% dilution can help with several mud properties and will allow the addition of calcium carbonate. The calcium carbonate adsorbs significant amounts of surfactant, which increases the filtrate/brine IFT. Pilot-test the additions to be sure the mud viscosity does not increase excessively. Also, wait until after the calcium carbonate pilot tests before measuring the IFT because the adsorption of surfactant on the solids increases the IFT value.

ES. A 2,000-V ES reading is not realistic in an OBM containing significant amounts of water. A 500-V reading may indicate a very good emulsion in an 80:20 oil/water-ratio fluid. Instead, focus on having enough emulsifier present so there is no observable free water in the mud filtrate during an HP/HT fluid-loss test.

5.4 Mud-Filtrate Tracers

5.4.1 Characteristics of a Good Tracer. Because the focus of LIC technology is obtaining data primarily available only from uninvaded rock, it is necessary to determine if (and how much) mud-filtrate reached the center of the core. Spiking the coring fluid with a unique chemical tracer that travels with the mud filtrate is essential to reliably quantifying filtrate invasion.

Some key characteristics of a good tracer include the following:

- The tracer must be soluble in the base fluid of the mud (i.e., water or oil) for it to travel with the phase of the filtrate in which the tracer is dissolved.
- It is preferred that the tracer be a chemical not found in the reservoir so that the filtrate is easily distinguished from the connate fluids. D_2O is an exception to this rule, but it is an effective tracer because its background level in reservoir brines is generally very low and constant. If the tracing chemical is present in the reservoir, the goal is to achieve at least a 10:1 contrast between the concentration in the mud filtrate and that found in the reservoir fluids.
- The tracer should be used in a high enough concentration to quantify mud-filtrate invasion as low as 1 to 2% PV. Dilution, in either the Dean-Stark analysis or from other extraction processes, makes higher initial tracer concentrations necessary.
- The tracer should be chemically stable under downhole conditions and throughout the core analysis measurements. This stability means the material should be thermally stable, should not be affected by the chemically reducing environment in a reservoir, and should not react or precipitate on contact with the reservoir rock, crude oil, or connate water.
- The tracer should be biologically stable (i.e., the material should not be a food source for bacteria). The tracer concentration needs to remain stable until all core analysis tests are completed.
- A good tracer should not be appreciably surface active (i.e., it should not adsorb on either the mud solids or on the reservoir rock). The tracer should travel freely with its phase of the mud filtrate and not react with its environment.
- The laboratory analysis for the tracer should be fairly simple and inexpensive.
- If possible, the tracer should be relatively inexpensive.
- The tracer should be as "green" as possible (i.e., it should not cause the mud or core fluids to result in negative environmental impacts or disposal problems).
- If possible, the tracer should be readily available either in the field or from the nearby chemical support infrastructure. Materials requiring long lead times can complicate LIC projects.
- A good tracer should be safe for rig personnel to handle. Safety information and safe handling procedures should accompany the tracer material when it is shipped to the field.
- The tracer should not evaporate faster than the mud's base fluid.
- For the degree and extent of mud-filtrate invasion to be accurately determined, the tracer concentration must be uniform throughout the mud system. It is helpful, but sometimes difficult, for tracer concentrations to be monitored at the rigsite throughout the coring period.

Finding a tracer that meets all of these criteria can be a challenge, but keeping the general principles in mind and identifying any places in the coring/core analysis process in which the data might be compromised maximize the chances of project success.

5.4.2 Adding Tracer to the Mud System. To accurately track the extent to which mud filtrate invaded the core, a tracer must be evenly distributed throughout the liquid phase of the mud. Because tracers are usually in limited supply at the rig (and sometimes expensive), they are most commonly added only after a successful displacement is achieved. In rare cases, a coring fluid badly contaminated during displacement may have to be replaced. A reliable way to achieve uniform tracer distribution is to first add it slowly and evenly over one full circulation of the mud system, and then to circulate the mud system at least twice more. The tracer should be added at a location with good mixing, perhaps by the agitator in the suction pit. All mud pits should be kept in active circulation to keep tracer concentration variations from developing.

5.4.3 WBM Tracers. A wide range of chemicals have been used as tracers in WBM over the years, including salts, alcohols, dyes, and radioactive materials. Many of these chemicals may help, crudely, to differentiate between mud filtrate and formation brine during a well test, but they are poorly suited to quantitatively track filtrate invasion into a core. For example, nitrate (usually supplied as ammonium nitrate fertilizer or sodium nitrate) is a common well-test tracer. Nitrate is a mild oxidizer and when circulated downhole into a chemically reducing environment (i.e., reservoir), it quickly begins to convert to nitrite and then to ammonia. Nitrate is also an excellent food for bacteria. This information all means that the tracer concentration decreases with time (at an unknown rate) and underpredicts the amount of mud-filtrate invasion. Lithium ions (and all other cations) also underpredict mud-filtrate invasion because they are consumed by ion exchange with the rock. Thiocyanate ion is thermally and chemically reactive. Dyes are surface active and adsorb onto reservoir rock. Alcohols tend to evaporate from the open mud pit.

The following tracers proved useful in LIC:

- ***Bromide.*** Bromide ions, usually supplied as sodium or potassium bromide salt, are a reasonably good coring tracer. Being an anion, bromide does not substantially ion exchange with the clays of the reservoir rock. Bromide is thermally and chemically stable, not a bacteria nutrient, and fairly easy and inexpensive to analyze by means of ion chromatography. Sodium bromide is usually readily available in the field because it is sometimes used in heavy completion brines. Potassium bromide needs to be purchased from a chemical supply house. To minimize costs, buy technical grade material in bulk quantities.

  Bromide has a few weaknesses as a tracer. It is uncommon in reservoir brines, but does occasionally occur, especially in higher total-salinity fluids. In Alaska, a number of North Slope reservoirs contain approximately 50 ppm bromide; and in the Middle East, some reservoir brines can contain up to 1,500 ppm. If the composition of the formation brine is known and the natural bromide concentration is low, then bromide is probably the most inexpensive, easiest-to-obtain tracer. If the bromide level is unknown, or high, then switching to iodide may be prudent. Another consideration for bromide (and all other salts) is that the core plugs must be crushed and extracted with water to recover all of the tracer. This procedure is undertaken after all the other tests on the core plugs are complete, but ultimately the core plugs are destroyed.

  The dose rate for bromide should be approximately 1.0 ppb for most muds, which results in approximately a 2,000-ppm bromide ion concentration.

  Any laboratory with ion chromatography capability should be able to perform the tracer analysis.

  When sending critical samples to an independent laboratory, a good idea is to always include some "blanks" (containing accurately known concentrations of the components to be measured) to check the quality of the reported data. Standards should be prepared in a base brine that matches the salinity of the field-analysis samples. Background salinity can affect measurement accuracy.

- ***Iodide.*** Iodide ions, supplied as sodium or potassium iodide salt, are an excellent tracer for coring. Iodide has most of the benefits of bromide but is usually not found in reservoir brines at levels above 1.0 ppm. There are occasional exceptions; therefore, always initially check the formation brine composition if possible. Iodide is typically five to six times more expensive than bromide, but the necessary treatment level is only 0.2 ppb for sodium iodide (approximately 500 ppm iodide). Again, any large chemical supply company should be able to supply this salt (bulk and technical grade). The analysis for iodide is also performed by ion chromatography. Standards should be prepared in a base brine that matches the salinity of the field brine samples. Again, background salinity can affect measurement accuracy. See Appendix E for sample processing and tracer calculations.
- ***D_2O.*** Deuterium is an isotope of hydrogen with one neutron. It is nonradioactive and nontoxic. Water ("heavy water") made with two deuterium atoms instead of two hydrogen atoms has a molecular weight of 20 instead of 18 and can, therefore, be easily measured using a mass spectrometer. Because its properties are very similar to those of water, D_2O is recovered from the core during the Dean-Stark extraction along with the connate water in the core. This recovery means that the core plugs do not have to be crushed and extracted as is required in the analyses for bromide and iodide salts. The laboratory provides the deuterium concentrations of the mud filtrate and the Dean-Stark water samples. With this information, the extent of filtrate invasion can be calculated.

  The difficulty with deuterium is that it is naturally present in water (including formation brine) at approximately 156 ± 10 ppm (Pallatt et al. 1991). This difficulty means that fairly high concentrations of deuterium must be added to a mud to be able to detect filtrate invasion down to 1 to 2% PV. A safe addition level is to add 850 ppm to the mud, based on the water-phase content of the mud system. Addition levels of 300 to 500 ppm are sometimes made, and some companies have successfully used a deuterium concentration as low as 150 ppm above background. Lower levels are less expensive, but data quality can suffer, especially when quantifying low mud-filtrate invasion levels.

  All chemical suppliers stock small containers of D_2O, but bulk-quantity sources are very limited.

  The material is high purity, with typically 99.3% of the hydrogen atoms being deuterium. The cost is substantial, running approximately USD 110 to 123/lbm (USD 240 to 270/kg) as of this writing (2010), with approximately 285 lbm (130 kg) needed per 1,000 bbl of mud (if adding 850 ppm). Also, because heavy water is used in nuclear research and development, some countries (e.g., those in the Middle East) may only be allowed to import small amounts per year. There are only a few manufacturers of D_2O, and the price is the lowest if purchased in bulk directly from a manufacturer.

  Laboratory assay for D_2O concentration costs approximately USD 40/sample as of this writing (2010). Once the laboratory analyses are complete, the mud-filtrate invasion calculations may be done as outlined in Appendix J.
- ***Tritium/Tritiated Water.*** Tritium is also an isotope of hydrogen, but including two neutrons. Unlike deuterium, tritium is radioactive, although it is a weak beta emitter and poses no danger once diluted into the mud system. Tritium does not naturally occur in reservoir connate brines. Only approximately 350 millicuries are needed [as water molecules containing one hydrogen atom and one tritium atom (HTO)]. This amount represents < 100 cm^3 of material needed to adequately provide tracer to 1,600 bbl of mud. The goal is to never exceed 50% of the maximum concentration allowed without special disposal considerations. The maximum amount permitted varies from country to country.

  The regulations concerning tritium are more extensive than its mild radioactivity would seem to justify. This difference led to a sharp decline in the use of tritium since the 1980s, despite its cost effectiveness (roughly USD 8,000/well). Only a service company licensed by the Atomic Energy Commission (in the USA) can supply tritiated water to the rigsite and mix it into the mud system. The details of using tritium on a project should be discussed with an experienced service company.

The laboratory analysis for tritium is performed by a process called liquid scintillation counting. Some of the water, including a proportional amount of the tritiated water recovered from the core plug during the Dean-Stark extraction, is carefully weighed into a vial of liquid scintillation cocktail. The radioactive decay of the tritium causes this cocktail to fluoresce in proportion to the amount of tritium present. The light from the fluorescence is easily quantified by the analytical equipment. The nature of this analysis means that if the sample is not clear and colorless, then data accuracy is somewhat reduced. The laboratory performing the tritium analyses is frequently the same company that supplies the tracer. Laboratory analysis costs are approximately USD 50 per sample.

5.4.4 OBM Tracers. Less research has been done by the industry on oil-phase tracers, but several good options are available.

Bromonaphthalene. 1-bromonaphthalene is soluble in oil, does not adsorb on reservoir rock, and has been found not to chemically degrade after 7 days at 350°F. Bromonaphthalene can be extracted from the core plugs by the flushing action of the toluene used in the Dean-Stark extraction, and then the toluene containing the bromonaphthalene is analyzed by way of a gas chromatograph with an electron capture detector. Appendix F contains an example test protocol for bromonaphthalene analysis. *Note: 1-iodonaphthalene also works well (i.e., 30 ppm is sufficient), but is no longer routinely manufactured and must be special ordered at approximately 10 times the cost of bromonaphthalene.*

Bromonaphthalene may be ordered from any of the major chemical-supply companies. Bulk bromonaphthalene costs approximately USD 75 to 80/kg, as of this writing (2010). A rough estimate for analysis costs is USD 75/sample for small numbers of samples (20 to 30), and approximately USD 50/sample for 100+ samples (all delivered at the same time).

The target concentration is 100 ppm in the oil phase of the mud. An example calculation is provided in Example 5.1.

Example 5.1 Assume the daily retort indicates that the mud contains 70 vol% oil. The density of this particular base oil is 6.83 lbm/gal, which means each 42-gal barrel of mud contains 200.8 lbm (91,082 g) of base oil. The target is 100 ppm; therefore, 100 millionths of 91,082 g is 9.11 g, which is the amount of tracer to add per barrel of mud. The density of bromonaphthalene is 1.487 g/cm^3; therefore, 9.11 g is equivalent to 6.13 cm^3. The result of these calculations varies depending on the density of a specific base oil and the actual oil content of the mud

Isotag. Isotag™ is a fairly recent development in tracers. For an OBM coring tracer, the company supplies a hydrocarbon where one or more of the hydrogen atoms have been replaced by deuterium. A typical treatment level is 2.2 lbm (1 kg) of Isotag™/400 bbl of mud. The cost is approximately USD 10/bbl. Sample analysis costs include a USD 3,000 setup fee and USD 175/sample, as of this writing (2010).

5.4.5 Mud Samples for Tracer Background Measurement. To determine the amount of mud filtrate invading the core, the background amount of tracer in the mud must be accurately known. To establish this background amount, periodic samples of the mud should be taken during the coring operations. The established practice is to take a time-lagged sample corresponding to the beginning of the core, at each 10-ft interval thereafter, and at the end of coring (i.e., four samples for a 30-ft core, or seven samples for a 60-ft core). A good sampling point is the mud pit below the shale shakers. The mud engineer usually takes the samples; but if that individual is busy, anyone in communication with the rig floor can take samples at the appropriate intervals. Taking accurately time-lagged samples is a challenge, especially if there is not a mud logger on the well to help track the time lag. Samples of 500 to 1000 cm^3 are adequate, unless other mud testing is planned. All samples should be clearly marked (i.e., with permanent ink) with the well name, date, core number, and depth. For WBMs, the sample containers should be tightly sealing plastic bottles (i.e., polyethylene Nalgene® bottles). Sealing the lids with a few wraps of electricians' tape can help prevent leaks. *Note: Do not use metal cans for WBM samples because they quickly rust and leak.* For OBMs, the situation is the opposite. Base oils can swell and soften plastic bottles; therefore, metal cans with screw lids must be used. Also, because OBM is combustible, the cans need to be US Department of Transportation (DOT) (or equivalent regulatory organization in other countries) approved for shipping. The mud samples should be shipped with the core to the core analysis laboratory. If wooden crates are being used to ship the 3-ft core sections, the best procedure is to combine all the mud samples in one separate crate for shipment.

5.4.6 Cost Considerations. Because tracer testing is fairly expensive, there are some cases in which it may not be necessary to analyze every core plug to determine the degree of mud-filtrate invasion. For example, if the core was cut at a fairly constant ROP and a few plugs from the highest-permeability portion of the core do not show filtrate invasion, then it can be reasonably inferred that mud-filtrate invasion did not occur in the lower-permeability portions of the core. Another option is to only test for tracer every 5 to 10 ft instead of every 1 ft. If these widely spaced samples show low mud-filtrate invasion, then it is reasonable not to analyze the other core plugs. If extremely detailed data are required, as in field equity-allocation determinations, then all samples should be analyzed.

5.4.7 Tracer Data. The contract tracer-analysis laboratory should provide for each sample a sample description, weight of sample on receipt, tracer concentration (ppm) from analysis, dilution factor, total tracer concentration (ppm), weight percent plugging fluid, and qualitative measure of base oil. The data are usually provided in digital format by the contract tracer-analysis laboratory to the contract core-analysis laboratory.

5.5 Mud Displacements

5.5.1 General. The goal of a displacement is to thoroughly displace, at a reasonable cost, the old mud out of the hole, leaving uncontaminated premium coring fluid in place. Clean displacements are usually achieved by pumping a series of spacers ahead of the coring fluid. Displacements can be very complex, but some rules of thumb are given as follows:

- Heavy fluids reasonably displace lighter fluids cleanly, but lighter fluids finger through heavier fluids and create a long interface. If the coring fluid is heavier than the drilling fluid, then the displacement is reasonably efficient, and the spacers should be designed to be heavier than the mud and slightly lighter than the coring fluid. If the coring fluid is less dense than the mud, then such an unfavorable displacement can be improved by making the spacers very large and viscous.
- Viscosity has an effect similar to that of density. Thick or high-viscosity fluids displace thin or low-viscosity fluids efficiently, but thin fluids finger through thick fluids and take a long time to displace the thick fluids out of a hole. Thinning the mud in the hole before the displacement helps, as well as making the spacer(s) large and viscous (at least the last one).
- For spacer size, a good rule of thumb is that a thorough displacement is obtained if the spacer is sufficient to fill 1,500 ft of hole. LIC most commonly occurs in an 8.5-in. hole, thereby translating to spacers of approximately 75 bbl each. Spacers this large may result in complications with respect to mud-pit management on some rigs.
- Regarding the number of spacers to use, occasionally a single spacer is adequate, but usually two spacers (1,500 ft each) result in a much more efficient displacement. The first spacer is designed to effectively mobilize the old mud. It is usually characterized by low viscosity (to thin the old mud) and contains surfactants or other chemicals to help with mud removal.

The second spacer is usually coring fluid with extra viscosifier added to cleanly sweep the first spacer and old mud out of the hole.

- The displacements should be made at high rates (10 to 12 bbl/min in an 8.5-in. hole, if possible) and pumped continuously. If pumping is stopped before the displacement is complete, gravity increases the length of the interfaces.
- Pipe rotation is generally thought to help keep the old mud mixed and to reduce dead spots. Rotating approximately 40 to 50 rev/min usually helps achieve a clean displacement. Reciprocation, on the other hand, is considered to lengthen the interfaces by plunging the tool joints back and forth through them. Assuming the hole is stable and has been circulated clean before the displacement, reciprocation is probably not needed.
- Determining when clean coring fluid is present at the end of a displacement is always a challenge. One simple approach is to add a sack of medium or coarse nutshell lost circulation material (LCM) to the last spacer. The appearance and tapering off of this material on the shaker screens serves as a clear indicator of the passing of the last spacer. Other useful indicators include pump-stroke counts, color, pH, density, and ES (for OBM). Even in high-angle holes, the displacement to quality coring fluid is usually complete by the end of the second spacer.
- It is fairly common for a core point to be a short distance below an intermediate casing shoe. If this is the case, the cement should be drilled and the casing shoe should be tested with the old mud to prevent contamination of the coring fluid.
- The shale shakers should be dressed with 50- to 80-mesh screens during the displacement and for the first circulation or two thereafter. Thick mud at spacer interfaces can blind fine-mesh screens, and partially hydrated polymer in the coring fluid can be screened out. Fine-mesh screens should be reinstalled before coring begins.
- The logistics of displacing an entire mud system can be quite complex depending on many issues related to mud delivery, rig plumbing, displacement rate, and mud haul-off logistics. There are too many variations to discuss here but the coring expert, drilling fluid engineer, and rig supervisor should prepare a detailed plan well in advance of the actual mud-displacement activities.

5.5.2 WBM to Oil-Based Coring Fluid. This displacement can usually be accomplished with two spacers. The goal is to minimize water contamination of the all-oil coring fluid (<2% water). The displacement can be aided by thinning the old mud before displacement, much like what is done before cementing.

Spacer 1. This spacer should be water-based, thin but viscous enough to support barite, approximately the same density as the coring fluid, and it should match the WBM with regard to salinity (i.e., type and amount) and pH. For an 8.5-in. hole, the volume should be approximately 75 bbl. This spacer can contain additional dispersants to help remove gelled mud.

Spacer 2. This spacer should be oil-based and quite viscous [yield point (YP) of 30 to 40]. The easiest way to accomplish these specifications is to add approximately 4 to 5 ppb of organophilic clay to 75 bbl of coring fluid. An effective way to monitor for the end of the displacement is to take ES readings at the "possum belly" (i.e., the mud-return tank on the shaker). The target value is 1,000 to 1,500 V, and this value should be achieved by either the end of the spacer or within the following 20 to 30 bbl. A sample with a 1,000-V ES should retort with < 10% water. The next 30 to 40 bbl of fluid may also be above the 2% water target, but this small amount of water is subsequently diluted within the entire mud system. Displacing until a 2,000-V reading is obtained results in wasting a considerable volume of expensive oil-based coring fluid.

5.5.3 WBM to Water-Based Coring Fluid. This displacement can usually be accomplished with two spacers. The goal is to prevent the contamination of the coring fluid with the surface-active agents and other components of the WBM. Thinning the WBM before displacement (as before cementing) improves the displacement efficiency.

Spacer 1. This water-based spacer should be thin, but able to support barite. It should have salinity and density similar to those of the coring fluid, have reasonable fluid-loss control (4-ppb starch, 1-ppb low-viscosity PAC), and contain biocide. The 75-bbl size is a reasonable size for an 8.5-in. hole. It is probably best if this spacer does not contain any dispersants or other surface-active agents. However, extra chemical may be warranted if the WBM being displaced either has high gel strengths or is considerably heavier than the coring fluid.

Spacer 2. This spacer should be approximately 75 bbl of coring fluid viscosified to a low-shear-rate viscosity (LSRV) of 40,000 to 60,000 cp with approximately 1 ppb of xanthan gum (determined by pilot test).

Note: Displacing from a freshwater WBM to a KCl WBM in open hole can rapidly destabilize water-sensitive shales and lead to wellbore instability caused by salinity shock. It is advisable to drill the entire hole section with a KCl-based mud if an openhole displacement is required.

5.5.4 OBM to Water-Based Coring Fluid. This type of mud displacement is uncommon because OBM is such a good coring fluid. However, displacing an OBM to a water-based coring fluid may be required if wettability is a key coring goal and the core is likely to be invaded because of a low ROP. A complete mud displacement is important because of the high-surfactant loading in most OBMs. Contamination of the water-based coring fluid with surfactants can compromise its ability to prevent alteration of reservoir wettability. OBMs are difficult to displace because they significantly oil-wet all surfaces, and because water contamination can cause an OBM emulsion to "flip," become water external, and highly viscous. In this type of mud displacement, three spacers may be more suitable than two.

Spacer 1. This spacer serves as a buffer between the OBM and any water-based fluids that follow. It should be an all-oil fluid, much like the coring fluid outlined in Section 5.3.1. This all-oil fluid can minimize the viscosity problems encountered when a regular OBM oil-external emulsion turns to a water-external emulsion. The spacer can be weighted with barite and needs no more than half of the fluid-loss material listed for the coring fluid. It should be loaded with strong emulsifiers/wetting agents to help it tolerate contamination with the water-based fluids that follow. It should be approximately the same density as the coring fluid and approximately the same viscosity as the OBM. If a heavier fluid is being displaced or the hole is high angle, the spacer should be large (75 bbl). Otherwise, approximately 40 bbl should be sufficient.

Spacer 2. This spacer should be water-based and loaded with water-wetting and oil-solubilizing surfactants (4 to 10% by volume). Citrus-oil-based solvents can also be helpful. The spacer should be fairly thin but able to support weighting material. The density and salinity should be similar to those of the coring fluid. It should contain some fluid-loss-control components (4 ppb starch and 1 ppb low-viscosity PAC) and a biocide. The volume of the spacer should be 75 bbl.

Spacer 3. This spacer should be very viscous, with an LSRV of 40,000 to 60,000 cp. The simplest approach is to take 75 bbl of the coring fluid and add approximately 1 ppb of xanthan gum polymer (determined by pilot test).

5.5.5 OBM to Oil-Based Coring Fluid. This displacement is also a fairly uncommon displacement, but an easy one. The goal is to keep the surfactants and water in the OBM from contaminating the coring fluid. A single spacer should be adequate. Use 75 to 100 bbl of coring fluid viscosified with an extra 4 to 5 ppb of organophilic clay (pilot test for YP of 30 to 40).

5.6 Solids Control

Extensive use of calcium carbonate complicates the normal use of solids-control equipment. Calcium carbonate (2.7 g/cm^3) has almost the same density as drilled solids (i.e., approximately 2.6 g/cm^3); therefore, hydrocyclones and centrifuges separate out both materials. Every effort should be made to run 200-mesh screens on the shakers (after displacement). This procedure may not be possible on some rigs; the use of all equipment should be optimized. Removing the maximum amount of solids on the first pass simplifies the maintenance of the coring fluid. The coring-fluid density may slowly increase because of the high-density slugs used when tripping the core out of the hole. Running the centrifugal equipment sparingly and evenly over one or two circulations of the mud can offset the density increase. Alternatively, mud companies can now supply very coarse barite for making slugs. The material is easily screened out when the slug is circulated to the surface. If clay content, viscosity, and mud weight continue to increase, dilutions can be made with unweighted new mud (4 ppb clay). Dilution should not be made if it can be delayed until coring is complete because it disrupts the homogeneous distribution of the tracer(s). Any mud additions must include tracer(s), and the mud should be circulated until the tracer(s) is evenly distributed (at least two full circulations of the mud system). Filling the mud pits as full as practical at the start of the job can frequently eliminate the need for any dilution. Approximately 50 to 75 B/D of whole mud is lost at the shakers, centrifuge, and other solids control equipment.

Chapter 6

Rigsite Operations

6.1 Overview

Rigsite operations are truly where "the rubber meets the road." Considerable preplanning for a low-invasion coring (LIC) project is critical, but there are always surprises at the rig. There are always on-site decisions that must be made immediately. The primary goal of achieving and preserving uninvaded core material must be kept in mind as changes and compromises are made.

Some rigsite questions to consider are as follows:

- Is all the equipment in place?
- Is the mud in good shape?
- Did the first bit perform as expected?
- Which bit should be run next?
- Was full core recovery achieved? If not, why not?
- What did the core look like? Was it in gauge? Was there any damage?
- Was torque a problem?
- Did the coring mud meet specifications?
- Was there any tripping damage (i.e., gas expansion, cracks) to the core?
- Was there any core-processing damage?
- Were core depths accurately recorded?
- Was the core plugged and preserved as rapidly as planned?
- Were all the rigsite operations safely executed?
- What can be done more rapidly during the next coring run?
- Were there enough people helping to complete the activities as rapidly as desired?
- Were all the mud samples taken as planned?
- Were quality-control core slices taken?
- Does anything need to be expedited to the rig?

This list could go on and on, but it serves to highlight the need to have experienced people at the rigsite who truly understand the technologies and can react appropriately.

6.2 Personnel

Several extra people are needed at the rigsite to make an LIC project proceed smoothly with a high probability of success. This need raises logistics issues such as

- Are there enough beds at the rig for the extra people?
- When should each of the team arrive at the rigsite?

Arriving a day or two before coring begins allows time to check equipment and prepare the core-processing area. Arriving too early is not an efficient use of everyone's time, but arriving early is far better than being late. It is always a guessing game, but there must be coordination of travel plans based on day-to-day updates from the rig supervisor and the rigsite geologist.

All people must be especially sensitive to the role of the rig supervisor, who bears the ultimate responsibility for everything that happens on the rig. The rig supervisor's input is needed on the coring plan, and early on a copy of the final document must be provided to him. Key considerations for the rig supervisor are as follows:

- The rig supervisor's help must be solicited with rig-specific issues.
- All rigsite activity must be cleared through the rig supervisor.
- Listen carefully to the rig supervisor's concerns.
- Keep the rig supervisor fully informed as the project progresses.

Be quick to accept appropriate responsibility if issues are raised from headquarters on items such as rig time or costs. The rig supervisor's understanding of the importance of the project is invaluable to a successful completion.

6.2.1 Coring Project Manager. This role should be filled by a single individual; sometimes a partner company also wants a representative at the rigsite during coring and core-handling operations. The coring project manager is in charge of preplanning, but should still arrive on site approximately 2 days before coring. This will allow time to check equipment and expedite any last-minute items, brief the rig crew on the details and importance of the project, oversee the layout of the core-processing area, hold safety meetings, and supervise the mud displacement operations. If full-scale core slabbing, plugging, and preserving are to be done on site, then a third preparation day may be appropriate. The coring project manager works with the other specialists and oversees all phases of the project, including bits and other coring equipment, coring mechanics, core lay-down, sampling, core processing and preservation, mud, tracers, and shipping.

6.2.2 Coring Engineers. There is sufficient work and enough hours of intense activity to justify having two coring engineers on LIC projects. At a minimum, the lead coring engineer should be a very experienced and a detail-oriented individual. The lead coring engineer should be involved in preparing the equipment list and should insist on new equipment whenever possible (i.e., inner-barrel connectors and catcher shoes). The lead coring engineer should arrange for adequate spares and tools. The coring engineer should arrive at the coring company's regional office a week or more before the job to carefully review all the equipment before it is shipped to the rig. Simple problems, such as mismatched threads, can cause serious complications. The focus is on maximizing ROP even if bit life is shortened. If poorly consolidated rock is expected, the lead coring engineer needs to be very experienced with either the Hydro-Lift™ or other full-closure catcher systems.

6.2.3 Rigsite Geologist. This individual is present at the rig during drilling to the core point. The rigsite geologist is responsible for picking the core point(s), helping keep the depths correct on all core pieces, and supplying preliminary core descriptions from the chips gathered at each 3-ft cut.

6.2.4 Coring-Fluid Specialist. This person is usually a drilling-fluid expert from the operator's drilling technology group. The coring-fluid specialist should have worked with the mud company well in advance to establish the proper formulation for the coring fluid and to pilot test the mud formulation when using local materials. Certain products may need as much as 6 to 8 weeks in transit to reach remote rig locations. The coring-fluid specialist may want to arrive as much as a week before coring begins to oversee the mud mixing and transportation to the rig. The coring-fluid specialist should oversee the mud-pit cleaning, mud displacement, tracer addition, mud sampling, solids control, and mud maintenance. The coring-fluid specialist also usually helps with the core processing.

6.2.5 Core-Processing Technicians. If the core is to be expedited to a laboratory for slabbing, plugging, and preservation, then two technicians can handle the rigsite work. If all these functions are to be completed on-site, then the total number of people needed is six to eight. The technicians should arrive approximately 2 days (3 days, if full processing is performed on site) before the first core arrives at the surface to have time to check equipment, take safety training, set up the core-processing area, and prelabel crates and sample bags. The main job for the technicians is conducting the core lay-down, mark-up, cutting, sampling, purging, capping, crating, cataloging, and the shipping process. The technicians usually oversee the logistics issues of preparing the core to be shipped out as soon as possible and alerting the commercial core-analysis laboratory as to its expected arrival time.

6.2.6 Roustabouts. Extra help is usually needed with cutting the core into sections and moving the crates. If two rig hands are available, the job proceeds more quickly and safely. If no roustabouts are available to help, then increase the number of core-processing technicians appropriately.

6.3 Coring Fluids—Displacement, Tracers, and Sampling

Coring fluids are covered in great detail in Chapter 5. Some of the key considerations for rigsite operations are as follows:

- OBMs are superior to WBMs for most jobs. WBMs are preferred if wettability/relative permeability are key goals, and if high ROP for low mud invasion is not ensured.
- Properly designed job-specific coring fluids are better than modified field muds.
- Include clay for a low-permeability base filter cake, adequate fine calcium carbonate for bridging, and considerable soft, deformable colloidal materials for fluid-loss control.
- Keep the coring fluid uncontaminated. Use clean mud pits, piping, and solids control equipment and clean transportation vessels.
- Degree of filtrate invasion cannot be quantified/tracked without including one or more mud-filtrate tracers.
- Design and perform the displacement to coring fluid carefully, using multiple large spacers.
- Do not begin coring until the coring fluid meets design specifications.
- Minimize changes in the coring fluid and tracer concentration until coring is completed.
- Take mud samples after every 10 ft of coring to determine the baseline for tracer concentration.

6.4 Equipment—Coring Contractor and Core-Analysis Contractor

Equipment lists vary from job to job, but the two following example lists are reasonable starting points. The examples assume

- An outside coring contractor
- An outside core-analysis contractor
- The core is slabbed/plugged in an off-site laboratory
- WBM is used
- The core is consolidated
- The hole from which a 4-in. core is being cut is 8.5 in. in diameter

6.4.1 Coring Contractor Equipment. This example uses Baker Hughes Inteq equipment.

- Specific core bits as dictated by lithology. Bracket the expected best bit(s) with backup bits suited for somewhat harder and somewhat softer rock. Prepare for sand/shale sequences. The bits must be antiwhirl with polished and reinforced cutters, face discharge, extended catcher shoe, no extra gauge cutters in the core-bit throat, and a total flow area (TFA) of approximately 1.0 in.^{2}. A 20° back-rake on the crown cutters is usually best, with 30° on the OD gauge cutters

Fig. 6.1—Core cradle (modified offshore basket) with shuck, inner core barrel, and chain clamps.

- HT-30 and HT-60 core barrels (6.75 × 4 in.). Heavy-duty core barrels can take more torque when operating at a high ROP. At least one complete back-up core barrel is needed. *Note: The heavy-duty assembly is stiffer and harder to work through doglegs.*
- (2) Mud motors, if needed. Mach 2XL or similar (latest generation).
- Drop-ball circulating sub. Allows the circulation of bottoms up without flowing through the core barrel and places LCM, if required.
- Aluminum inner core barrels. Perhaps 30% more length of inner core barrels on-site than the total planned core footage.
- Elevator subs
- Crossover subs, as needed
- End protectors
- $8^{7}/_{16}$-in. integral stabilization at 30-ft increments (for 8.5-in. coring assembly)
- Connections compatible with bottomhole assembly (BHA)
- Cutting cradle (see Fig. 6.1). Offshore basket rigged with rollers and chain clamps (three to four) to support core/inner core barrel while it is being marked and cut into 3-ft sections. Approximately five rollers are needed. If a shuck (a rigid protective sleeve) is being used, then three are needed (3-ft spacing) near the end where the cutting takes place. If a shuck is not used, all the crossbar/roller assemblies can have chain clamps for the inner core barrel to be cut from both ends at the same time. For unconsolidated cores, the entire 30-ft core section needs to be frozen before cutting into 3-ft sections.
- A shuck for core barrel lay-down is frequently made of 7-in. casing. The purpose is to keep the inner core barrel from bending (i.e., breaking core) while being transferred from the rig floor to the core-processing area. The shuck can be slotted to lie over the rollers on the cutting cradle and allow the inner core barrel to easily slide out. The shuck should have lift-eyes cut into the top and a drain hole (or several) cut at the bottom (see **Figs. 6.2 through 6.5**).

Fig. 6.2—Slot in shuck to accommodate rollers on core cradle.

Fig. 6.3—Core shucks with slots for core cradle rollers and "ears" for pick-up.

Fig. 6.4—Shuck lay-down with wheeled boot.

Fig. 6.5—Close-up of wheeled boot for shuck lay-down.

- (8) Caps with lift-eyes to screw onto the inner core barrels for lifting sections
- (2) Cordless electric drills for tightening hose clamps holding rubber caps on inner core barrel 3-ft sections. Hex sockets work best, but heavy standard screwdriver bits work also. Must have rechargers compatible with rig/camp power.
- Pressure-relief valves for inner core barrel or 1/16-in. bits for drilling gas vent holes
- A 15-ft long cable with lift-eyes at both ends for laying down inner core barrels
- (3) Rigid, four-wheel, 180° cut, heavy-duty pipe cutters for cutting aluminum inner core barrels **(Fig. 6.6)**
- (12) Spare cutter wheels
- Pneumatic chop saw
- (3) Saw blades, diamond
- Inner-barrel rubber end caps. Two end caps for each 3-ft section, with 30% excess (4.75-in. diameter).
- Stainless-steel hose clamps for attaching rubber caps. Two hose clamps for each 3-ft section, with approximately 30% excess.
- Duct tape to seal any air leaks from under hose-clamped rubber caps (1 roll per 30 ft of core)
- O-ring, upper (Viton™)
- O-ring, lower (Viton™)
- Complete swivel assembly
- Long distance (LD) adjusting system, complete
- Split shim set
- Cartridge cap
- Cartridge plug
- (5) Thrust bearings
- (2) Bearing retainers
- Inner-barrel plug
- (4) Drop balls
- (3) Pressure-relief plugs
- (4) Top-end shoes
- (4) Extended pilot shoes (low-invasion); (one if single core)
- (6) Spring catchers (one, if single core)
- Outer core, barrel-extension sub for jam indicator
- (6 new) Rotating sleeve, inner-barrel stabilizers. Modified window type.
- (20) Machine screws for window cover
- (2) Outer core, barrel-extension/slick subs
- Sliding-sleeve, drop-ball, circulating sub or equivalent tool complete with steel and plastic balls

Fig. 6.6—Cutting inner core barrel with four-wheeled pipe cutter.

- (5) Redress kits for circulating sub
- Float subs for core barrel, 2 ft, 2.5-in. inside diameter (ID), 6.75-in. OD, threads compatible with BHA
- Core tong shoe
- Core tong handle
- (2) Pipe wrenches, 36 in.
- (4) Inner-barrel clamps (standard)
- (2) Chain tongs
- Bit breaker, HT-30 (high torque)
- (2) Outer core, barrel-repair subs, 1 ft long
- (2) Outer-core, barrel-repair subs, 2 ft long
- (3) Inner-barrel clamps (modified with wedge for core lay-down)
- Guillotine shear-plate basket
- (2) Inner-barrel, capped-thread protectors
- (2) Ball pick-up tools
- (2) Core markers
- (6) Inner-tube centralizer caps
- (4) Inner-tube clamps
- (8) Lift rings for inner core barrels
- (1) Pedestal
- (2) Pressure-relief plugs
- (1) Top end modified for pedestal
- (1) Collet core catcher, inner-tube collet assembly
- (2) Inner-tube shoe, pilot type, standard core catcher
- (1) Inner-tube top end (long), standard core catcher
- (1) Inner-tube top end (short)
- (1) Standard core catcher, spring type
- (4) Thrust bearings
- (2) Outer-tube repair sub, 4 ft long
- (2) Crossover subs (for core barrel); determine drill collar connection
- (10) Inner-tube centralizers with clamps
- *Note: Additional equipment needed for Hydro-Lift™*

6.4.2 Core-Analysis Contractor Equipment. This example uses Core Labs equipment.

- (1 large, commercial roll) Aluminum foil
- (2) Cordless screwdrivers, heavy-duty, rechargeable with rig voltage, with extra bits for attaching lids to shipping crates
- Boxes for transporting two 3-ft sections of core (expected +30%), heavy-duty, interior spaces each 8 × 40 in., lined with foam carpet underpad or similar material, heavy-rope handles
- Chop saw for cutting QC samples, extra diamond blades
- (2 rolls) Duct tape
- (2 dozen) Ear plugs
- (6 rolls) Glass-strapping tape
- (3 dozen pairs) Gloves, cloth
- Hard hats, safety glasses, steel-toed boots, and other safety gear for each person
- (2) Large chisels
- (1 dozen of each color) Lumber crayons, yellow and white
- (1 dozen of each color) Permanent markers, waterproof, red and black

- (3 per core) Plastic bottles (e.g., Nalgene™) for water-based mud samples, 1 L, water-tight, screw-top closure, or metal containers (1 gal) for oil-based coring mud samples with screw-top closure
- Nitrogen bottle with safety stand, regulator, and needle valve for purging inner core barrel sections, 4 ft of 0.125-in. steel tubing, and 20 ft of 0.25-in. nylon tubing
- Tubing cutter
- (4) Purging nozzles, 3.5 ft, with 15 ft of plastic hose and wire probe for cleaning plugged nozzles
- (2) Note pads and pens/pencils
- (2) Nut drivers and bits
- (2 to 3 per 3-ft core section) Plastic strips to fill inner core barrel space and prevent core motion, 1/16 × 1 × 36 in.
- (Approximately 5 per core) polyethylene, resealable plastic bags for core pieces in catcher shoe and inner-barrel stabilizers
- (1 roll) Laminate and a heat sealer to make pouches for QC samples (described subsequently in this chapter); wrapping and wax sealing also an option
- (2 large boxes) Rags, cotton
- (2) Rock hammers or small sledge hammers (3 lbm)
- (2 large, commercial rolls) Plastic wrap
- Tape measure (100 ft)
- Polyethylene core tubing (50 ft)
- Dust masks and dust collection system

6.5 Strapping and Drifting the Pipe

When pulling out of the hole to pick up the first core assembly, carefully strap (i.e., measure) the drillpipe to determine an accurate depth for the top of the first core. Report the depth to the coring project manager. Drift the pipe [i.e., run a "rabbit" (an internal-drift-diameter gauge) larger than the biggest drop ball planned through the pipe] to be sure all ball activations can take place.

6.6 Making Up the Core Barrel

Carefully check the drilling bit for damage as it comes out of the hole. Any junk left in the hole quickly damages a light-set, low-invasion coring bit. If necessary, make a preliminary cleanout run with a junk basket before cutting the first core.

Making up the core barrel is the responsibility of the coring engineer. Listed subsequently are a few comments based on cutting a 60-ft long, 8.5-in. diameter core through consolidated rock. Choices, such as the initial bit type and core barrel length, must be determined ahead of time.

6.6.1 Coring BHA. ***Outer Core Barrel.*** This example assumes BHI equipment and a 60-ft coring assembly.

- Crossover sub, if needed
- Float sub (to prevent debris from entering barrel and plugging ball seats)
- Safety joint pin
- Safety joint box
- Long distance (LD) adjusting sub
- Stabilizer, 4 ft to $8^{15}/_{32}$ in.
- Outer core barrel, 6.75-in. OD × 5.375-in. ID
- Stabilizer, 4 ft to $8^{15}/_{32}$ in.

- Outer core barrel, 6.75-in. OD × 5.375-in. ID
- Stabilizer, 4 ft to $8^{15}/_{32}$ in.
- Bits

Inner Core Barrel.

- Thrust bearing
- Aluminum inner core barrel
- Inner-barrel stabilizer/connector

Lift the completed coring assembly high enough to check the spacing between the end of the catcher shoe and the back of the bit. This spacing needs to be precise. The inner core barrel has a greater coefficient of heat expansion than the steel outer core barrel. Sufficient space must be available to allow for this differential expansion. If the length of the inner core barrel expands too much, it contacts the back of the lip at the entrance to the bit, potentially causing the inner core barrel to turn with the outer core barrel, and resulting in core jamming/damage. If the gap between the inner core barrel and bit is too large, an excessive amount of coring fluid exits the assembly in the core-bit throat, and the extent of mud-filtrate invasion increases. The coring engineer has a table to define the proper gap based on the expected downhole mud-circulating temperature.

If a mud motor (or any other small clearance tool) is being run, then the normal process of dropping a ball cannot be used to divert the mud flow from the throat of the core barrel to the annulus between the inner and outer core barrels. The ball can be run in a temporary holder (e.g., BHI's DAFD tool) where it can be released by high pressure/flow. This preserves the ability to circulate at mud rates high enough to clean fill (i.e., settled rock debris) from the bottom of the hole before the start of coring. Circulating through the inner core barrel also ensures any material pushed into the inner core barrel while tripping in is flushed away.

Also, one should flow test the mud motor with it suspended in the upper part of the wellbore before tripping it all the way to bottom. If there is a problem with the motor, then a flow test will determine the problem and save the expensive rig time used to trip the pipe all the way to the bottom of the hole and back out again to change the defective motor. During these flow tests, several pump rates (200, 250, and 300 gal/min) should be used.

Second Coring Run. The bit performance should be examined, and, if necessary, changes should be made to increase ROP. Because ROP is critical, do not rerun moderately worn bits. Save them for any later non-low-invasion cores.

6.7 Tripping In

The coring assembly is quite stiff and may encounter resistance in tight curvatures/dog-legs. If necessary, gently circulate and ream through these areas. When using WBM, it is possible for the inner core barrel to become jammed with softened shale from the wall of the hole as the assembly is pushed through a tight interval. Washing through tight intervals reduces this risk.

Trip in the last stand slowly, watching for fill. Wash through any fill to bottom and circulate out of the wellbore.

Drop the ball and watch for the circulating mud-pressure increase that indicates the ball has seated and diverted mud flow from the inner core barrel to the annulus between the inner and outer core barrel. Note times and pressures.

6.8 Coring Mechanics

The goal is to maximize ROP to minimize mud-filtrate invasion. The experience and training of the coring engineer guides this optimization.

Cutting the first few feet of core should be done carefully until the bit has established a stable cutting pattern on the bottom of the hole. Low rev/min (perhaps 60), and only 1,000- to 2,000-lbm WOB is the usual approach. Rev/min and WOB can be ramped up and balanced, as necessary, to maximize ROP, within the constraints of torque limit and bit life. In most sandstones, low-invasion-style bits usually only require approximately 4,000 to 6,000 lbm to achieve high ROP.

On a foot-by-foot basis, record the depth, ROP, WOB, rev/min, torque, circulating rate, and circulating pressure. When the core barrel is full, pull up to break off the core. Tag the bottom of the hole, and circulate bottoms up before coming out of the hole.

6.9 Tripping Out

When using OBM, supply disposable, oil-resistant suits for the coring engineers and floor hands to minimize skin exposure to OBM.

Use automated breakout equipment, and do not rotate the drillstring on the way out of the hole. Set the slips gently to avoid jarring the core. *Note: Sudden drillstring movements can occasionally cause the core to fall out of the core barrel.*

Rapid gas expansion can damage a core; therefore, the tripping rate needs to be slowed as the core reaches a depth at which free gas begins to evolve, particularly the last few hundred feet when much of the gas expansion and flow (as a result of pressure reduction) occurs. For consolidated rock, use a pulling time of 3 minutes per stand from 1,500 ft to 500 ft, and 6 minutes per stand from 500 ft to the surface. *Note: These times are pulling times, not including connections times.* Gas-expansion damage can be significant for weak or unconsolidated cores. See Chapter 7, which discusses coring poorly consolidated sands for which even slower trip rates are required.

If hydrogen sulfide (H_2S) is potentially present in the core, use an H_2S sensor to "sniff" at each connection starting at the top of the BHA. Also sniff at the inner-barrel connector when the window cover is removed to break the core at the 30-ft mark. If significant levels (i.e., greater than 10 ppm) of H_2S are detected, then the core processing may need to proceed with personnel using supplied-air systems. Based on limited experience, H_2S seems to evolve quickly from the core while tripping out, but this serious safety issue must be planned for far in advance. If measurable H_2S levels persist, the inner core barrel can be flushed with nitrogen using a rubber end cap, tubing, and a nitrogen bottle.

When the coring assembly reaches the surface, check the mouth of the bit to see if there is core in the core catcher.

6.10 Core Lay-Down

The core must be transferred from its vertical position in the derrick to a horizontal position in the core-processing area. During this maneuver, there is the risk of flexing the inner core barrel and cracking the core inside—sometimes in several places. The easiest way to avoid this problem is to slide each 30-ft section of inner core barrel into a rigid protective sleeve or "shuck," usually a 30-ft section of 7-in. casing (see Fig. 6.6). The description that follows assumes the presence of a shuck. However, if a shuck is not available, then the core can be lowered with minimal flexing by using two nylon slings, each placed 7.5 ft from the ends of the 30-ft inner-barrel section, as shown in **Fig. 6.7.**

- Lift the inner-barrel shuck to the drill floor, and place it in the mousehole.
- Break the safety joint thread on the outer core barrel.
- Lift the inner core barrel to expose the catcher at the bottom. Slightly loosen the catcher connection. Because the catcher has a firm grip on the core, excessive turning of the catcher can induce a spiral fracture in the core as much as 8 ft long. Run the inner core barrel back into the outer core barrel.
- Place the inner-barrel clamp at the top of the top section.
- Break out the safety joint from the inner core barrel.

Fig. 6.7—Core transfer to work area with the core properly supported to prevent flexing and breakage.

- Attach thread protectors with lift eyes to the safety joint and to the top of the inner core barrel. Connect the two lift eyes with a 15-ft cable using shackles.
- Remove the inner-barrel clamp, and lift the inner core barrel using the elevators until the inner-barrel bearing (connector) is exposed.
- Place the inner-barrel clamp at the top of the second (lower) section of the inner core barrel.
- Loosen the bottom thread on the inner-barrel bearing (connector).
- Remove the window from the inner-barrel stabilizer by removing the four set screws.
- Check for H_2S with a sniffer, if necessary.
- Drive the clamp-with-wedge assembly into the core through the window.
- Complete unscrewing the inner core barrels at the bottom thread of the inner-barrel stabilizer.
- Screw a thread protector onto the bottom pin of the inner-barrel stabilizer, and then remove the clamp-with-wedge assembly.
- Place the top section of the inner core barrel gently into the shuck. Lift the shuck with an air winch. Without any banging or sudden shocks, slide the shuck down the V-door to the pipe rack. A wheeled boot (Fig. 6.5) can make the transition smoother.
- If available, a crane can be used to lift the shuck and place it on the inner core barrel processing rack (see Fig. 6.1).
- Use a chisel to remove any core interfering with the threads on the top of the lower inner core barrel section. Screw on a lift eye, and connect the inner core barrel to the safety joint with a 15-ft cable.
- Place the lower section of the inner core barrel gently into the shuck. Lift the shuck with an air winch. Without any banging or sudden shocks, slide the shuck down the V-door to the pipe rack. A wheeled-boot (see Fig. 6.5) can make the transition smoother.
- If available, a crane can be used to lift the shuck and place it on the inner core barrel processing rack (see Fig. 6.1).

6.11 Core Processing

6.11.1 Core Processing for Rapid Transfer to the Laboratory. Usually, the most cost-efficient approach is to process the core only into 3-ft capped sections of the inner core barrel and then ship them to the analysis laboratory. Because diffusion of mud filtrate and tracer toward the center of the core continues, the core should arrive at the laboratory as quickly as possible to that it can be fully processed within 48 hours of reaching the surface.

- Slide the inner core barrel out the shuck onto the core-processing rack.
- Clean the inner core barrel thoroughly for the depth markings to stick.
- Locate the top of the core (broom handle or equivalent), and mark the inner core barrel at this point with the depth determined from strapping the pipe out the hole.
- Using waterproof permanent ink markers, place red and black orientation stripes on the inner core barrel following the convention "red on right, looking uphole."
- Mark each foot of the core on the inner core barrel, and write the depth a few inches below the mark.
- Put an X by every third foot mark to indicate where the cuts should be made to process the core into 3-ft lengths.
- Sequentially number each segment starting at the top of the core, and also mark the core number.
- Securely clamp the inner core barrel down using the chain clamps on the processing rack.
- Cut the core into 3-ft lengths using 4-wheel, 180° pipe cutters. Once the aluminum barrel is cut, lightly score the core with the pipe cutter. This usually produces a clean break. If needed, use a hammer and chisel to break the core.
- The geologist may want to take chips for a preliminary core description at each 3-ft cut.
- Cap one end of the core segment using a rubber cap (4.75 in.), and seal with several wraps of duct tape and a stainless-steel hose clamp.
- Slide a 1- to 2-in. wide, 3-ft long plastic strip into the annulus between the core and the inner core barrel to keep the core from moving or breaking during shipping.
- Purge the air (in particular, oxygen) from the annulus between the core and the inner core barrel using nitrogen delivered through a hose and a 4-ft long, 1/8-in. diameter steel tubing assembly. Oxygen can cause wettability changes within the core. Be sure the high-pressure nitrogen bottle is safely secured to prevent it from falling over.
- Quickly cap/clamp the remaining end of the 3-ft core segment. If the core sticks slightly out of the inner core barrel and the rubber cap must be stretched on, then use duct tape and two hose clamps to ensure the cap does not come off during shipping.
- Place the core segments into wooden shipping crates marked with the company name, well name, date, depths, core number, core segment number, "fragile" stickers, and shipping address. The crates should be very sturdy with comfortable rope or wooden handles. The crates can be designed to hold either a single 3-ft section (approximately 80 lbm total) or two 3-ft sections (approximately 150 lbm) divided by a wooden partition. The interior space of the crate for each 3-ft section should be 40 in. long by 6 in. wide and should be padded with either foam carpet padding or equivalent. Undersized crates lead to pinched fingers and skinned knuckles for personnel.
- The core-analysis laboratory personnel should prepare a careful tabulation of the core segments.
- Mud/tracer samples should be shipped along with the core, but placed into a separate crate.
- The catcher assembly must be removed from the last inner core barrel section. Do not remove the catcher before cutting the core into 3-ft sections. Because the core is jammed in the catcher, trying to unscrew it can torque the core hard enough to induce spiral fractures up to 8 ft long.

- The core recovered from the catcher needs to be orientation striped and depth marked. A spare piece of aluminum inner core barrel can be used, or the pieces can be sealed in a polyethylene bag. The small pieces recovered from the inner-barrel connector areas also need to be oriented, bagged, and depth marked.

6.11.2 Slabbing, Plugging, and Preservation On Site. If the core cannot be shipped to a laboratory for slabbing, plugging, and preservation within 48 hours of reaching the surface, then this work must be done at the rigsite. The required equipment and procedures are described in Chapter 8.

6.12 Quality-Control Sampling

Because tracer diffusion continues while the core is being shipped to the laboratory, it is prudent to take an occasional QC sample to establish an invasion/saturation baseline **(Fig. 6.8).**

Every 30 ft, push a few inches of the core out the inner core barrel and make two saw cuts on the exposed end with a table-mounted diamond saw to obtain a 2-in. segment of the whole core. Do not take this 2-in. sample from near the end of the core or near any fractures (i.e., sites of greater mud-filtrate invasion). Saw the 2-in. disk into pieces, as shown in Fig. 6.8, and immediately preserve the numbered pieces separately in heat-sealed laminate pouches. Replace the remaining core and sample remnants into the aluminum inner-barrel section, filling the space created by the missing center section of rock with plastic wrap. Place plastic strips as spacers between the core and the aluminum inner-barrel section on one side of the core to prevent movement and core breakage. Purge the sleeve with nitrogen, and cap and clamp both ends.

6.13 Shipping to Core-Analysis Laboratory

The object of LIC is to obtain core plugs free of mud-filtrate invasion so that accurate in-situ water saturations, wettabilities, and brine compositions can be measured. When an LIC reaches the surface, the mud filtrate in the outer edges of the core continues to move toward the center of the core because of imbibition by capillary forces and diffusion. Core plugs must quickly be cut from the center of the core to avoid mud-filtrate contamination by these forces. The quickest way to slab the core and cut and preserve the plugs is to take all the necessary equipment and personnel

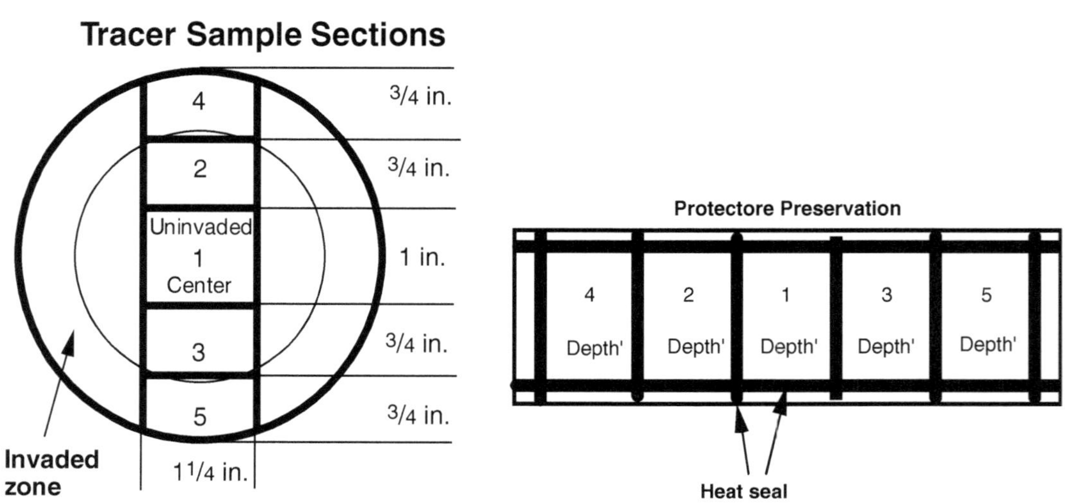

Fig. 6.8—Template for core sectioning and preservation of quality control (QC) samples.

to the rigsite. However, this is quite expensive and can be logistically unworkable, especially when there are space limitations. Field experience and engineering calculations indicate that if the core is slabbed and plugged within 48 hours of reaching the surface, then these sources of post-coring mud-filtrate contamination are avoided under most circumstances. This can allow rigsite costs and problems to be minimized by arranging for the core to be expedited to the nearest commercial core-analysis laboratory.

Various methods for rapid shipment of the core to a laboratory are possible. If the drilling operation is offshore, the options are moving the core either by helicopter or by workboat. In many cases, the workboat option is too slow, and shipping the core by helicopter is the only viable option. This means determining the weight limit the helicopter can carry and probably temporarily removing the passenger seats. Crates containing 3 ft of core weigh approximately 80 lbm each, and crates containing 6 ft of core weigh approximately 150 lbm. However, these numbers should only be used for planning purposes. A representative sample of the crates should be weighed on a job-by-job basis to obtain a more accurate estimate of the helicopter payload. Care should be taken to give the rig supervisor as much lead time as possible when estimating the total amount of time that the helicopter is needed. Scheduling may be a complicating factor.

A log should be made detailing the contents of each crate. Each crate lid should be labeled with the well name, core number, core depth, and laboratory address.

Once the core reaches the helibase/shorebase, the process becomes similar to transporting the core from a land rig. The core shipment must be met and trucked either to a local laboratory or to the airport for further transport. Air schedules must be determined well in advance to minimize delays. Adequate freight space on the plane must be reserved as far in advance as possible. Careful investigation must be done to determine the type of paperwork needed for sending the cores by airfreight and, if required, through customs. The core must be met on the other end by laboratory personnel for transfer to the laboratory. Care must be taken to avoid temperature extremes and to minimize temperature variations. Freezing is known to damage the core by several physical mechanisms. Excessive heat can cause fluids to evaporate from the core.

Transferring the core to the core-analysis laboratory is usually the responsibility of the contract laboratory, but oversight and assistance from the coring project manager are essential. The emphasis needs to be on speed and minimizing any physical damage to the core or alteration of its fluid saturations.

Once the core reaches the laboratory, processing should begin immediately (regardless of time of day or night) and should continue until all core plugging and plug preservation are completed.

In some cases, there is neither room on the rig for processing nor a way to get the core quickly to the laboratory at a reasonable cost. In these cases, a temporary core-processing facility (e.g., near the helibase) should be established.

Some of the items that must be considered are as follows:

- Slab saws
- Dust-handling system
- Trim saws
- Cutting fluid and fluid-delivery system
- Plugging machines (drill presses)
- Storage areas for supplies and cores
- Wax sealant dipping vat
- Drying rack for dipped samples
- Sufficient bench and table layout space
- Adequate lighting
- Air conditioning for the workers if possible, several fans if not
- Ample electricity supply

Chapter 7

Coring Poorly Consolidated Sandstone

7.1 Overview

Achieving full recovery with low mud-filtrate invasion is very difficult in poorly consolidated rocks. Poorly consolidated sandstones tend to fluidize and wash away instead of filling the inner core barrel. Even if the rock does make it into the core barrel, the unconsolidated sand grains can be substantially rearranged. While advancements have been made, there is no set of procedures that consistently guarantees high core recovery of poorly consolidated sandstones. The subsequent recommendations substantially improve the probability of high core recovery, but coring such rock is still an art. This chapter focuses on core recovery, because until recovery levels are sufficiently high, there is no point in pursuing low mud-filtrate-invasion considerations. However, most of the coring techniques subsequently described are compatible with achieving low mud-filtrate invasion.

7.2 Core Length and Diameter

The best approach when coring poorly consolidated rock is to cut short cores (as short as 10 ft in the poorest-quality rock). If 30-ft core barrels are successfully filled, then increasing the core barrel length to 60-ft core barrels can be considered. Ninety-foot core barrels are usually not attempted in poorly consolidated sandstones. As the length of the core barrel increases, so does the weight of the core pushing down on the new rock being cored at the bit. At some point, the rock at the bit begins to crush and wash away. In addition to poor core recovery, evidence of the onset of core damage includes the lack of an annulus between the core and the inner core barrel near the bottom of the core.

Use the largest core barrel the hole size allows. Cores less than 4-in. in diameter are very difficult to recover in poorly consolidated rock. BHI offers two sizes for their full-closure Hydro-Lift™ barrel: 6.75×4 in. and the HT-60 9.5×5.25-in. system.

7.3 Coring Fluids

The LIC fluids outlined in Chapter 5 are excellent choices for coring poorly consolidated rock even if low mud-filtrate invasion is not a goal of the project. The primary goal is to hold the core together long enough to get it up into the protection of the inner core barrel. A low-spurt-loss coring fluid quickly develops a low-permeability filter cake that gives the mud overbalance pressure something to push against—helping to hold the core together. Without a high-quality filter cake and sufficient overbalance pressure to hold the core together, the core is frequently fluidized into separated individual grains and, even if recovered, is of little use in understanding the rock sedimentary structures at the original reservoir conditions. Chapter 5 covers low-spurt-loss coring fluids in detail.

7.4 Mud Overbalance Pressure

For consolidated-rock low mud-filtrate-invasion coring, the standard recommendation is to use the minimum mud overbalance pressure consistent with safe drilling operations because high mud overbalance pressure results in a greater driving force to flow mud filtrate into the core. With poorly consolidated rock, the priorities are different. To achieve adequate core recovery, the mud and mud overbalance pressure must artificially hold the core together in its native state until it is safely pushed into the inner core barrel. Calculations, laboratory tests, and field experience have shown that 400+ psi overbalance pressure substantially increases the probability of high core recovery. Because poorly consolidated rocks are frequently at shallow depths and mechanically weak, it can sometimes be difficult to use 400+ psi overbalance pressure without risking lost circulation. However, low-spurt-loss muds can help reduce mud-loss problems, even under high-overbalance conditions. Fracture-gradient information coupled with equivalent circulating density estimates should be used to determine the highest-acceptable mud weight.

7.5 Bit Selection and ROP

In most respects, low mud-filtrate-invasion coring bits are better suited for coring poorly consolidated rock than standard bits. The reduced fluid flow through the throat of an LIC bit lowers the chance of the rock being fluidized. Most LIC bits have a face/throat discharge ratio of approximately 85:15 (vs. 100% throat discharge in old-style bits). In addition, the extended core-catcher shoe ensures the core is protected within approximately 1 in. after being cut. However, these bits are designed to maximize ROP and can actually cut poorly consolidated rock so fast that there is no time for an adequate filter cake to form on the core surfaces. The lack of this filter cake makes it very likely that the core will be washed away or fluidized and packed into the inner core barrel in random order. Building an adequate mud filter cake can be helped by choosing an LIC bit with a less aggressive cutting structure. For example, an eight-bladed, Genesis™ profile, BHC408 cuts slower than a six-bladed, Genesis™ profile, BHC406. The majority of BHI "Genesis BHC" bits incorporate extra cutters just behind the main row on each blade that can limit the depth of cut on each core bit rotation and improve observed WOB in very weak rocks. Unlike previous product lines, nearly all face-discharge BHC designs have been used to core poorly consolidated rock. Similar design features are found on the DBS FC264LI and FC284LI bits. To give adequate time for a filter cake to form, the ROP when coring unconsolidated rock should not exceed 60 ft/hr. Controlling ROP can be difficult in soft rock because the coring engineer does not observe consistent torque and weight on the bit. Using either the BHC406 or even a nine-bladed BHC409 (normally used for harder rock) can reduce the efficiency of coring to the point that clear WOB is indicated while still achieving 60 ft/hr.

The first-generation LIC bits still allowed approximately 40% of the coring fluid to exit at the throat; therefore, a new design has been developed. A new BHI BHC406 was designed with poorly consolidated rock in mind. The flow through the throat has been reduced to 15%, which is just enough to ensure that cuttings are cleaned away. The bit has six blades, larger face-discharge ports, and a somewhat parabolic profile. The track record with this bit has been excellent. Because of the reduced flow through the throat, the latest low-invasion BHC design can, and needs to, use more flow than normally used on unconsolidated rock. The minimum flow rate for the BHI BHC406 is approximately 100 gal/min, and up to 200 gal/min may be needed in sand/shale sequences.

7.6 Flow Rate and Rotary Speed

Even with the right coring bit and coring fluid, unconsolidated rock can be disaggregated by a high mud flow rate. Mud flow rates of 40 to 80 gal/min are common with normal low mud-invasion bits (higher with the latest low-invasion BHC design). Rotary speed should also be reduced to help limit ROP. Rotation rates of 40 to 75 rev/min are common. (For example, at Hamaca field in Venezuela, rotation rates of 60 to 75 rev/min with minimal WOB gave a low cut per cutter and better core recovery.) The combination of mud flow rate and rev/min should be adequate to clean the bit, assuming the coring rate is still 60 ft/hr.

7.7 Full-Closure Core Barrel

Poorly consolidated rock is too weak to be held inside the inner core barrel by a standard, split-ring core catcher. A full-closure barrel is required. When coring is completed, a second ball is dropped that diverts mud flow to lift the entire inner core barrel slightly; therefore, a pair of powerfully spring-loaded, clamshell-type catchers can slam together and seal the core in the barrel. There is even a full-closure system available that employs additional mud flow rate and pressure to crush a device that looks like an aluminum soda can, so the bottom of the assembly is sealed. It is an innovative concept, but it tends to activate early if the circulating mud pressure spikes, and it does not usually completely seal if there is rock in the throat of the "can."

Full-closure systems should be backed up by a secondary standard, split-ring core catcher. If the coring run ends in a few feet of harder rock, the standard core catcher ensures full core recovery. In fact, when cutting sand/shale sequences, it is a reasonable policy to try to terminate coring with a few feet of shale across the catcher. However, this approach may not work in extremely weak sands. The extra vibration generated in transitioning from sand to shale—and while cutting shale—is thought to damage the soft sediment in the inner core barrel. Remember, damage to the core can be visualized by way of CT scans.

7.8 Tripping Rates

Even if poorly consolidated rock is successfully cored, tripping the core out the hole too quickly can damage the core's integrity. Because the rock is inherently weak, it can easily be pushed apart by the breakout and expansion of free gas as the core is tripped to shallower depths (i.e., lower pressures). Where consolidated core has approximately a 0.25-in. annulus with the inner core barrel, it is common for unconsolidated core to be pushed apart by gas expansion until it packs the inner core barrel to the point where it cannot be removed without splitting the inner core barrel. CT scans sometimes show spider-web-like cracks even when the core has not obviously expanded. The loss of the original grain structure makes it difficult to estimate the true in-situ rock properties.

Tripping rates should be slow enough to allow time for the gas to flow out the core without pushing the grains apart. If the oil pressure/volume/temperature (PVT) behavior is known, then bubble-point calculations can be made to predict the depth-vs.-pressure and define when solution gas will be liberated. Reduced tripping speeds should begin at least a few hundred feet before the bubble-point pressure is reached. If the gas composition is not known, then default tripping times can be used. These recommended tripping speeds are the actual pulling times per stand without including connection time.

2,500 to 1,500 ft	2 min/stand
1,500 to 600 ft	4 min/stand
600 ft to surface	10 min/stand

If these tripping rates still produce core damaged by gas expansion [i.e., no annulus between core and inner core barrel, or fractured/rubblized core as shown by CT scans (note that visual observation of slabbed surfaces does not show most damage)], then trip the next core even slower. If the oil has a high viscosity, then extremely slow rates may be required to give sufficient time for gas to devolve from the oil and flow (gas relative permeability is very low). For viscous oils, it may be impractical to pull slow enough to prevent gas-expansion damage.

In addition to slow tripping rates, using either check valves or drilled holes to vent gas from the core can reduce gas-related problems. If parts of the core expand and trap gas in the inner core barrel, the core can be violently expelled when the inner core barrel is uncapped. A typical approach is to use approximately one check valve per foot with 30° phasing. Some people have experienced situations in which the check valves are prone to plug, preventing fluid drainage at the surface. Small vent holes ($^1/_{16}$ to ⅛ in.) have been used successfully without plugging and without appearing to allow excessive amounts of mud to circulate inside the inner core barrel (no observable damage to core).

7.9 Core Stabilization

Poorly consolidated core should be laid down very gently (i.e., no bumps, bending, or rotation). It must also be stabilized before further processing. Cutting the barrel into 3-ft sections, capping them, and shipping them are likely to lead to changes in the sand-grain structures, loss of their original configuration, and potential mixing with mud. The preferred approach is to freeze the core in dry ice (i.e., solid CO_2); but if the location is remote and dry ice is not available, then the core can be stabilized by injecting a fixative-like epoxy or plaster of Paris into the annulus between the core and inner core barrel. The entire 30-ft section should be stabilized before it is cut into 3-ft sections. The 3-ft sections should be very carefully padded and immobilized in the shipping crate.

7.9.1 Freezing. If possible, each 30-ft section of core should be frozen whole. The core-analysis contractor normally supplies the dry ice, freezing boxes, and insulated shipping crates. Once frozen, the core can be cut into 3-ft sections, sampled, capped, and clamped. The sealed sections should then be placed in an insulated chest with more dry ice to keep it frozen until sampling can be completed. The 3-ft core sections should be firmly packed to prevent movement during shipping.

Note: Use dry ice in either snow or pellet form, because blocks of dry ice produce uneven freezing, especially with fiberglass inner core barrels, and may lead to redistribution of pore fluids and/or core damage. If only the block form of dry ice is available, break the blocks into small chunks (i.e., 2 in. or smaller).

Freezing is the preferred method of stabilizing unconsolidated core but is not a perfect approach. Dry ice, and also liquid nitrogen, can be hard to acquire in many remote locations. Pore water expands upon freezing, which may induce fractures or cause pore fluids movement. Also, the salts in the connate water migrate from the freezing water into the yet-to-be-frozen water within the core, which leads to higher salinities at the center of the core and lower salinities at the outer edges. Still, freezing induces fewer problems than the other options, such as epoxy or plaster of Paris (discussed subsequently).

7.9.2 Epoxy. As stated previously, stabilizing an entire 30-ft section of core is preferred (segmenting the core into 3-ft sections is too likely to cause rearrangement of the sediments). For example, Core Lab offers the CorSta™ process. Once the 30-ft section of core is gently transferred to the core-processing area, a series of holes are drilled along the top of the inner core barrel and at 135° down each side. The lower holes allow most of the mud to drain, which can be enhanced by very gentle air pressure (i.e., 0.5 psi) through the top holes. With the mud drained, epoxy is gently injected through the lower holes. When epoxy is seen at a top port, that hole is plugged for epoxy to be forced evenly throughout the inner core barrel. Once the epoxy is fully cured, the core can be cut into 3-ft sections.

The older epoxy procedure was as follows:

- Mark the barrel with the usual red-and-black orientation lines (red on right, looking uphole) and foot-by-foot depths.
- Carefully cut the inner core barrel into 3-ft lengths, minimizing vibrations.
- Support each 3-ft section as it is cut free, not allowing it to tip down or rotate.
- Obtain samples (spoon) from the upper end of each tube, and describe the annular space on the core summary sheet (e.g., note whether there are any signs of gas expansion damage).
- Seal the inner-barrel section with rubber caps and hose clamps.
- Seal any gas vents with duct tape.

- Keeping the orientation lines up (no rotation), place the 3-ft sections on a rack tilted no more than 30°, and cut holes (approximately 1-in. diameter) into the bottom of the rubber caps to allow mud to drain.
- Cover the bottom hole with duct tape. Rotate the upper cap (not the core) for the hole to be up, and pour epoxy into the inner-barrel section until the core is completely covered and the end spaces are filled. This step may mean making the tubes vertical for the final filling.
- Do not move the core until the epoxy is completely cured.
- Cover the top hole with duct tape.
- Very carefully pack the 3-ft sections into a shipping crate, so they cannot move during shipping (spray cans of settable foam work well). Note: The core is still vulnerable to damage.
- Make sure everyone knows to minimize bumping or jarring the crate during transport to the core-analysis laboratory.

The problems with epoxy are as follows:

- Epoxy heats to approximately 130°F while hardening.
- Epoxy can invade, to some extent, into cracks and into the sand matrix.
- Epoxy will set slowly at cool temperatures.
- If injected, epoxy can rearrange sediments from their original grain structure.

7.9.3 Plaster of Paris. This material can be used in a fashion similar to epoxy, but it has a slower set-up time, is somewhat brittle/fragile, can absorb water from the core, and may be harder to inject because of high viscosity.

7.10 Personnel

Coring, handling, sampling, and processing poorly consolidated rock are not for inexperienced coring engineers or core-processing personnel. If at all possible, employ an engineer who is experienced with full-closure-barrels and familiar with the best coring practices in the project area. Also, use core-processing personnel who understand the importance of having the core arrive at the contract core-analysis laboratory in a condition as close as possible to the native state.

Chapter 8

Core Analysis

This chapter discusses aspects of core handling, sampling, preservation, and basic core analysis procedures. Other documents have been prepared by industry experts covering these topics in considerably more detail, particularly API's *Recommended Practices for Core Analysis* (API *RP-40, Second edition,* 1998). Also, the contract core-analysis laboratory typically has standard procedures that it follows for these measurements. If an operator employs core-analysis specialists, it is preferred that they supply their own protocols, or at least review and approve those recommended by the contract laboratory.

8.1 Operator Representative at Laboratory

Even if the core acquisition phase has gone well, low mud-filtrate-invasion core data can be compromised in the contract laboratory if careful QC/QA is not maintained. One of the best ways to ensure high-quality laboratory work is to have an experienced person from the operating company at the contract laboratory to oversee all operations. Optimally, this person has a core-analysis and/or geology background. In fact, this person's job should begin several months before the core arrives at the laboratory, when an audit is conducted to identify any necessary procedure changes or equipment upgrades. The operator's core-analysis specialist and the lead core-analysis-laboratory representative should be involved with writing the core-analyses procedures document. Very detailed step-by-step protocols should be prepared, and QC/QA crosschecks—like logging sample weights at each step to track for possible fluid evaporation—should be used wherever possible.

The operator representative must be available at all hours because the core processing should begin the moment the core is received at the laboratory to minimize the time for further mud-filtrate contamination by diffusion and capillary forces.

8.2 CT

CT scanning the capped inner-barrel sections can determine whether the core is in good condition or disaggregated.

8.3 Core Spectral Gamma Ray Logging

Conduct spectral gamma ray measurements while the core is still preserved in the aluminum inner-barrel sections. Lay out the capped inner-barrel sections in depth order, butted tightly up against each other. Calibrate the core gamma ray unit with a 200 API gamma ray emitter before each core is run, or at least at the beginning of each day. Feed the core through at a constant rate. Digitize and plot the gamma ray response.

8.4 Core Slabbing, Plugging, and Preservation Protocols

Processing should begin immediately when the core is received in the laboratory and should continue until it is completed. The tracer and mud-filtrate continue to redistribute toward the center of the core because of saturation and concentration gradients. Tracer diffusion is fairly slow, but after several days (perhaps as long as a week), it becomes significant. In a WBM system, these effects cause the calculated connate water saturation to be too low and the calculated in-place hydrocarbon volumes and reserves to be too high when a sample is really only slightly invaded by WBM filtrate. Even if diffusion occurs, samples with no tracer in them at the time of analysis will yield reliable connate water saturations. Also, because the core is only preserved in capped inner-barrel sections, water can be lost from the core as vapor during daily heating and cooling cycles.

To minimize water evaporation, each inner-barrel section should not be opened until it is time for its core layout, marking, and slabbing. Only approximately 6 to 9 ft of core should be laid out at a time. Once the core is removed from the inner-barrel sections, it should be wiped with dry or oil-wetted rags to remove mud, fitted together, oriented with the apparent up-dip at the top of the layout tables, and depth marked and double striped for "upline" and slabbing orientation. The double-striped upline should be perpendicular to the bedding plane where the core is oriented to show maximum dip. Depths should be marked below the corresponding foot line. Orientation lines should be marked "white on the right" and "yellow on the left" (looking uphole) with waterproof markers, such as lumber crayons. Permanent markers (i.e., red and black, red on right looking uphole) should be used on a foil-wrapped preserved core.

Sampling and plugging protocols should be adapted to the needs of a specific project. What follows is a typical arrangement, but far from the only choice as to the number, size, and type of core plugs.

Note: It may be desirable to preserve some whole core (not slabbed) for future SCAL studies. However, continued diffusion of mud filtrate might alter saturations and wettability. It may be preferable to drill, trim, and preserve future SCAL plugs at this point in the core processing.

8.4.1 Slabbing. Core slabbing should be conducted as soon as the gamma ray measurements are completed. The core pieces should be fitted together and oriented for the same side to always be represented in the slab. Slab the core ¼ and ¾ to show the maximum bedding dip in the slabbed surface. Cross-bedding can complicate the process. The ¾-core slab should retain the yellow and white stripes and the depth markings made upon layout. Rotate the core 90° from these double-stripe markings and cut the slab. Slabbing should be conducted using cold nitrogen gas to cool the saw blade. If cold nitrogen gas is not available, a light mineral oil can be used as the cutting fluid. On the cylindrical surface of the 1-in. slab, double-stripe and depth-mark for core orientation. For frozen cores, use CT scans to determine bedding dip.

8.4.2 Routine Sampling Program. For routine samples, a 4-in. long piece (its top at the foot mark) is marked and sawed to separate it from the core.

8.4.3 Preservation. The ¾-core butt end of the core is immediately preserved in plastic wrap, aluminum foil, and two dips of core sealant. The ¼-core slab should be placed, in depth sequence, in a slab box padded with expanded polystyrene packaging inserts.

The ¾-core butt of the core should be placed on the bench with the top always to the left to help ensure the proper core orientation is maintained. Continue using this protocol when the core is transferred to the wrapping station. The wrapping stations should be covered with oil absorbent sheeting or some similar material to provide a soft surface that does not damage the plastic wrap and foil.

Wrapping of the core should begin as soon as possible after it has been slabbed. All of the ¾-core butt should be wrapped with six layers of plastic wrap **(Fig. 8.1)** and then three layers of aluminum foil **(Fig. 8.2).** The wrapped core should then be depth marked using white stick-on labels and a black permanent marker. The package should also be given red-and-black orientation stripes (red on right looking uphole).

The routine samples (4-in. pieces) should be preserved separately from the other core samples in a given foot. The same protocol for plastic wrap, aluminum foil, and depth-and-orientation markings should be followed. These pieces are used for cutting plugs as time and manpower permit.

The next step is to coat the wrapped ¾-core butt with core sealant. The coating with core sealant should be completed on all samples in 3 to 4 hours, except for samples awaiting core plugs. No

Fig. 8.1—Wrapping in six layers of plastic wrap.

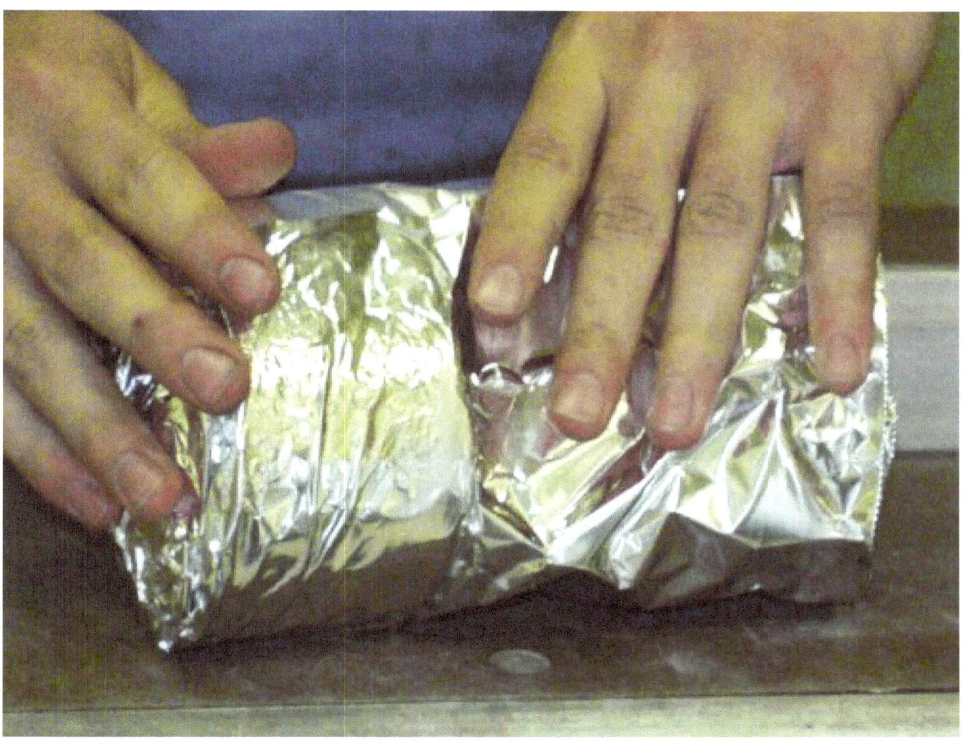

Fig. 8.2—Wrapping in three layers of aluminum foil.

samples should remain without core sealant protection for longer than 8 hours. Baling wire should be wrapped around the core sample as a handle for use during the dipping process. To prevent puncture of the aluminum foil, a few wraps of glass strapping tape should be wound around the middle of the sample where the wire handle is attached. Carefully monitor the core sealant temperature in the dipping vat to ensure the recommended temperature for the sealant is not exceeded. For example, CoreSeal™-type material should not exceed 330°F. *Note: If the sealant becomes discolored from being overheated, it should not be used and must be replaced.* The core should be completely submerged in the dip tank for 1 to 2 seconds and then hung on the cooling rack **(Fig. 8.3).** After the sealant has cooled for 3 to 5 minutes, the sample should be redipped to ensure a complete seal and to help protect the core from damage during shipping. After the second coat of sealant has cooled, cut off the wire just above the attachment point and seal that area with additional sealant spooned onto the spot. If this area is not sealed, a path is available for fluids to slowly evaporate.

Once all of the ¾-core butt pieces are preserved and placed in open core boxes, they should be checked for proper preservation and markings. Occasionally, the core sealant is too thick and the markings cannot be read. Then, the information has to be duplicated on the sealant coating on the outside of the sample.

8.4.4 Cutting and Trimming Core Plugs. It is common for the routine (RCAL) horizontal core plugs to be cut from the slab at 1-ft intervals over the entire length of core. To avoid any sampling bias, RCAL core plugs should be taken at a fixed distance from each foot mark (i.e., 0.2 ft below). There are usually no lithology guidelines for locating RCAL core plugs. Cut 1.5-in. diameter horizontal plugs parallel to bedding plane (i.e., perpendicular to the slabbed surface) using humidified

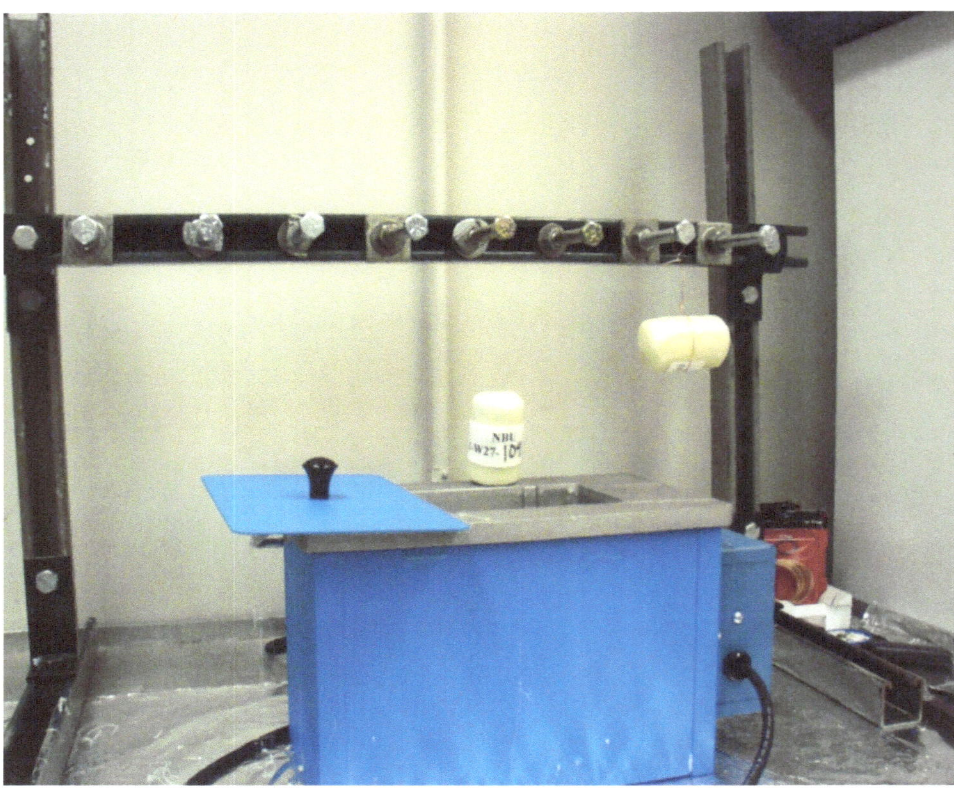

Fig. 8.3—Preserving in core sealant.

nitrogen gas as a coolant and cutting fluid. A light mineral oil can be used as the cutting fluid if oil contamination is not a concern. The core plugs should be cut completely through the middle of the slab to maximize plug length. When using humidified nitrogen gas, use nitrogen from a nitrogen cylinder; do not use gaseous nitrogen from a liquid-nitrogen bottle. Dust masks and dust-collector systems are required for the cutting of RCAL core plugs with gas.

The RCAL core plugs should be cut as early as possible in the core-processing sequence. Core processing, at least through cutting RCAL plugs, should be completed as quickly as possible, using shift work if necessary. **Figs. 8.4 and 8.5** illustrate the slab and horizontal plug locations.

If vertical RCAL core plugs are to be cut, they are cut from a 4-in. sample (see Fig. 8.4). Care must be taken to ensure the vertical core plugs are cut perpendicular to the bedding planes. After the RCAL horizontal plugs are cut, both corners of selected 4-in. pieces (perhaps one per 5 ft of measured depth) can be trimmed parallel to the bedding planes to give a flat surface for acquiring the vertical core plug (commonly 1-in. diameter by 1.5-in. length) as shown in the lower right diagram of Fig. 8.4.

Vertical RCAL Plug Location. Trim one end of the core plug with a dry saw to make the final plug 1.5 in. long. If the plugging bit is not perpendicular to the slab face, first trim the slab side of the core plug 1/8 in. (the width of the saw blade), and then trim off the end piece to a resulting core plug of 1.5-in. length. Mark the core plug number on the core plug and end piece. Preserve the core plug in plastic wrap (six layers) and aluminum foil (three layers). Relabel the sample by writing the core plug number on the foil. Place the wrapped core plug in a preweighed bottle; then seal, tape, and weigh the sample when preservation is completed. Label the bottles with well name, depth, core plug number, tare, and total weight. Wrap the end piece in plastic wrap, tape it closed, place it in a

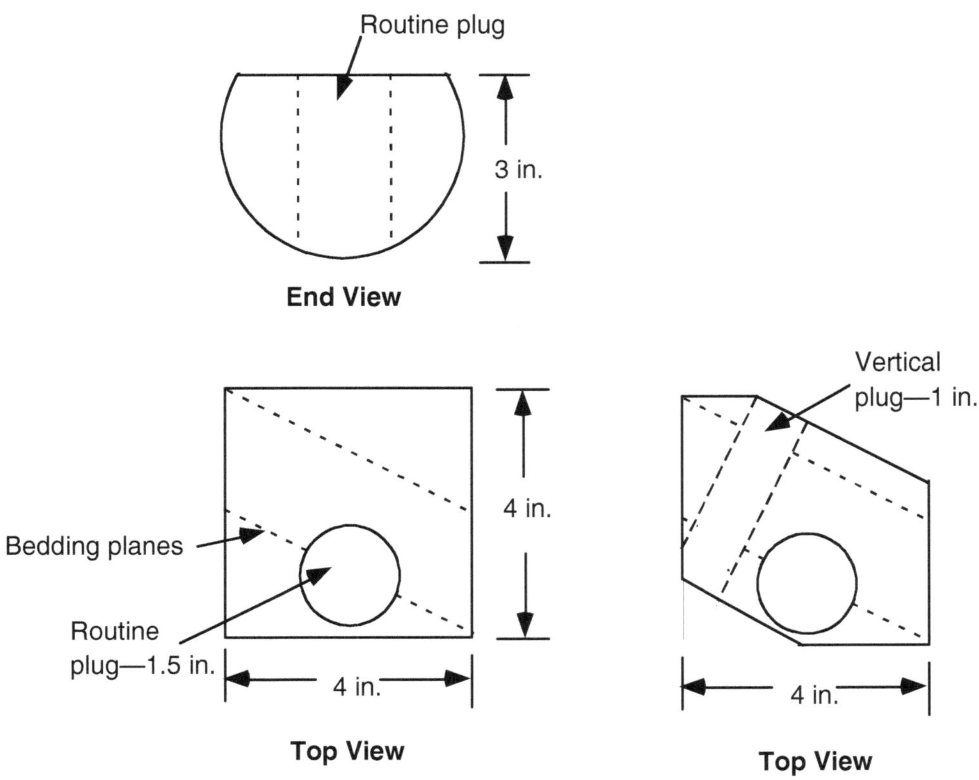

Fig. 8.4—Diagram showing core slab locations for cutting routine horizontal and vertical core plugs.

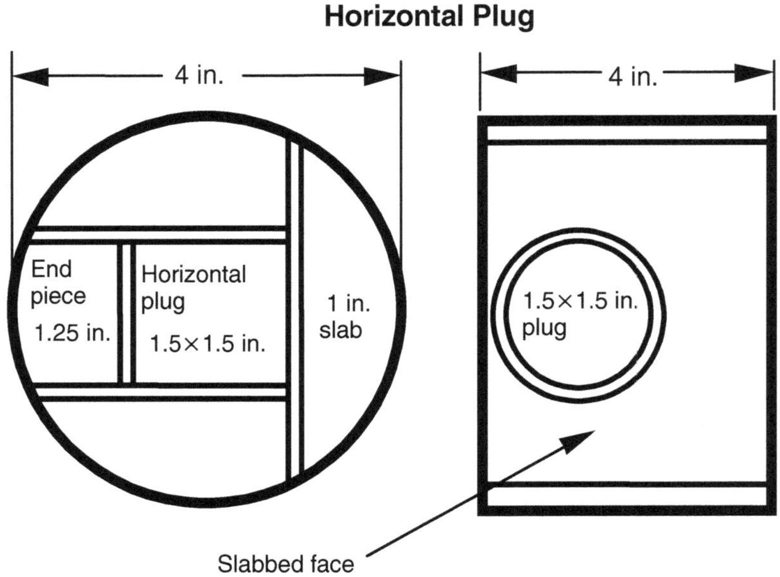

Fig. 8.5—Template for cutting horizontal core plugs.

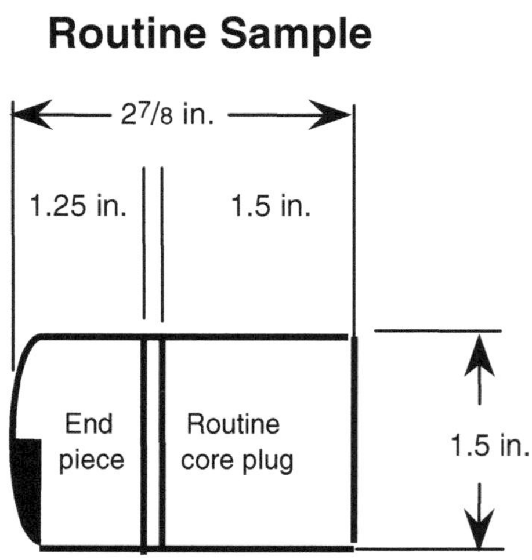

Fig. 8.6—Dimensions and trimming of routine core plugs.

plastic resealable bag, and mark the depth and core plug number on the package. Indicate end piece with "EP" following the core plug number (i.e., "10320-EP"). Fig. 8.5 and **Fig. 8.6** show the physical relationship of the core plug and its trimmed end.

RCAL Plug Ends. Clean selected RCAL core plug end pieces of hydrocarbons and salt; then dry and send for thin-section preparation.

8.4.5 Cutting and Trimming SCAL Core Plugs. Cut 1.5-in. diameter by 2.0-in. length horizontal SCAL core plugs parallel to the bedding plane, perpendicular to the slabbed surface, positioned in the middle of the slabbed face, and cut completely through the slab (see **Fig. 8.7**).

Use humidified nitrogen gas as the cutting fluid. (*Note: Do not use the very cold nitrogen gas off the top of the liquid-nitrogen tank*). If oil contamination is not a concern, a light mineral oil can be

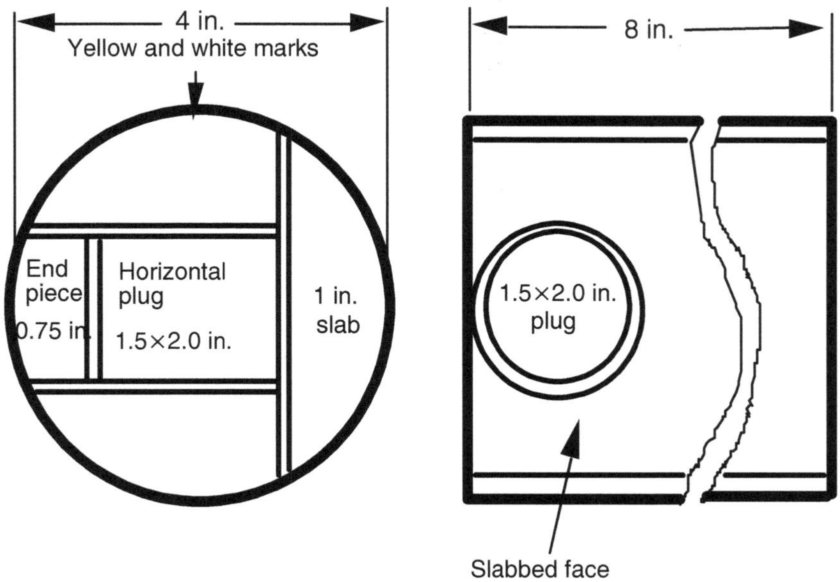

Fig. 8.7—Dimensions and location of core plugs for SCAL tests.

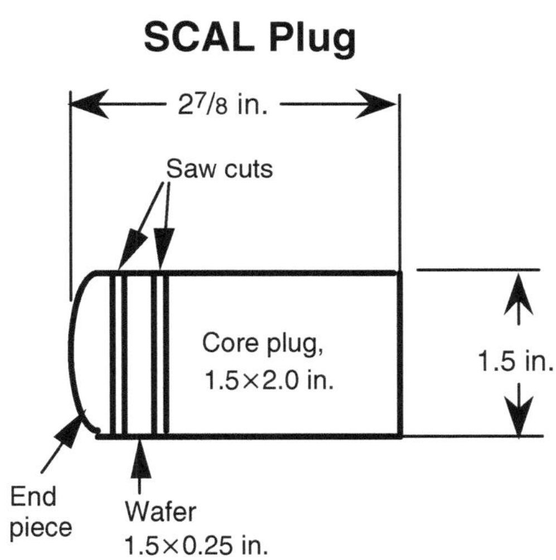

Fig. 8.8—Trimming SCAL core plugs.

used. Cut all the core plugs that can possibly be obtained down the center of the 8-in. core section (usually four core plugs).

Trim the end of the core plugs with a dry, double-bladed trim saw to make the plugs 2.0 in. long and exactly in the center of the whole core where the mud-filtrate invasion is minimized. Cut a ¼-in. wafer from the end pieces for the wafers to be directly adjacent to the core plugs. The relative locations of the core plug, wafer, and end piece are shown in **Fig. 8.8.**

8.4.6 Preservation of SCAL Core Plugs and Whole-Core Remnant. Identify the core plug number on the core plug, wafer, and end piece. Wrap the core plug tightly in six layers of plastic wrap, excluding air as completely as possible. This wrapping must occur on a soft, padded surface covered either with Oil-Sorbant™ tissue or similar material. Carefully fold over the excess plastic wrap on the ends; then, tape the ends closed with glass tape wrapped around the circumference at the center of the core. Wrap in three layers of aluminum foil. Wrap glass tape around center of the core plug, and then coat with double layers of core sealant. Measure and record the weight of each preserved core plug as a baseline to monitor any subsequent weight loss. Heat-seal the preserved core plugs in laminate pouches.

Wrap the end piece and wafer separately in plastic wrap, tape closed, and heat-seal the end piece and wafer in a laminate pouch.

Use a single section of laminate with three pockets—one for the core plug, one for the end piece, and one for the wafer. Label the complete package with depth, core plug number, and total weight. Prepare a summary listing of the core plug numbers, depths, and total weights.

The SCAL plugging consumes all the rock in the 8-in. interval; therefore, the core remnant does not need to be represerved. The SCAL core plugs may be analyzed in the same contract laboratory a different contract laboratory, or they may be shipped to the operator's technology center for specialized work.

8.5 Core-Analysis Procedures

8.5.1 Permeability to Oil. For cores from an oil reservoir, the permeability to oil should be measured under increasing net-effective overburden stress, up to the maximum stress level estimated at reservoir conditions (i.e., mean net-effective overburden stress). The data at the higher stress levels are used to develop a transform to adjust the routine air permeability data to in-situ reservoir-permeability values. These tests, performed before any other measurements, can be run on RCAL horizontal core plugs (perhaps 1 in every 5 or 10). The measurements can be made using refined oil previously tested and found to be compatible with the reservoir crude. It must not precipitate asphaltenes or cause damage to the rock matrix. A gradual shift from crude oil to refined oil is preferred with the core and injected fluids heated above the wax appearance temperature of the crude oil. Flush 3 pore volumes (PV) of a filtered 50:50 mix of crude oil and refined oil before injecting 3 PV of refined oil. Allow to cool to ambient temperature, and measure oil permeability. Kerosene is frequently a good choice to avoid compatibility problems. No more than 6 PV of this oil is injected for this measurement because large volumes of flow solubilize small amounts of water and remove that water from the core plug. Additionally, there is the possibility of fines migration. Measurements of pressure and flow rate need to be taken after each PV on injection until stable values are obtained. Later, after all RCAL core-plug analyses are completed on these plugs, they can be resaturated with 3% KCl brine, and then the permeability to brine can be measured.

8.5.2 Dean-Stark Water Extraction. The Dean-Stark water-extraction procedure is used to determine water saturation. If either an oil-soluble tracer or a water-soluble tracer like deuterium or tritium oxide were used, it is recovered from the core plug during the Dean-Stark water extraction. Fresh commercial-grade toluene must be used for each core plug analyzed to avoid cross-contamination. *Note: New containers of toluene must be sampled and analyzed for water contamination before use; water contamination up to 400 ppm is common.* The flasks and glassware must be thoroughly cleaned before each use to ensure there is no residual tracer or water present. Rinse the condenser, core chamber, and flask five times with 20 cm^3 of fresh toluene and then discard the rinses. Weigh a clean, dry Dean-Stark flask. Add toluene (approximately 100 cm^3) and preboil. Check the side arm for water and transfer the toluene in the side arm back into the flask. Dry the side arm. To minimize toluene losses, keep the system closed, except when either adding or removing the core-plug sample.

Remove the core plug from its preservation materials. The core plug is then weighed to an accuracy of ±0.001 g, placed in a redried glass or ceramic thimble to capture lost grains and reweighed to the same precision. If the core plugs are soft or poorly consolidated, then wrap them in thread seal tape, nickel foil, and screens before the last weighing. Obtain individual weights on the thread seal tape, foil, and screens. Then, without further delay, load the core plug into the Dean-Stark apparatus. The apparatus is then sealed. All tasks should be performed on one core plug at a time. *Note: Glass or ceramic thimbles must be used and not cellulose thimbles for the Dean-Stark water extractions because cellulose thimbles are hygroscopic and rapidly adsorb water from the laboratory air. That adsorbed water is later captured during the Dean-Stark water extraction and leads to overestimates of the water saturation in the core plug.*

Toluene is refluxed for a minimum extraction time of 48 hours and until no further change in the water volume in the receiving tube is observed after an additional 4 hours. Use a wire with a Teflon® swab on the end to scrape off water droplets from the neck of the condenser that fall into the receiving tube. Do not use soap. This procedure should be done at least 8 hours before the Dean-Stark water extraction is terminated for the collection of any additional water in the condenser to be observed. Record the volume of water to within ±0.01 cm^3. Remove the water from the receiver with a tared syringe, and then weigh the water. Remove the core plug, and weigh it to ±0.001 g. (When these samples have permeability measured on them, be aware that the Dean-Stark water extraction procedure may have resulted in broken fragile fibrous illite clays, if present, leading to artificially high measured permeability values.)

If an oil-based tracer is used, remove the flask and its contents from the Dean-Stark apparatus. Remove the toluene from the side arm with a syringe, and pour it back into the flask. Weigh the flask plus the toluene mixture. Using a funnel previously rinsed five times with toluene, pour the flask contents into a storage bottle. Be sure the cap is tightly sealed so it does not loosen during shipping. In addition, seal the bottle with tape. Label the bottle with well name, core number, core depth, and core plug number. These samples are then shipped to the contract tracer-analysis laboratory. The packaging should be consistent with US Department of Transportation (DOT) regulations or other government regulations.

8.5.3 Cleaning and Drying. The core plug and thimble should next be placed in a second piece of Dean-Stark glassware for extraction with 100% chloroform. The chloroform interferes with oil-based tracer analysis if chloroform is present in the toluene extract in a high concentration. Glassware used for Dean-Stark water extractions should never be used for chloroform extraction until all the Dean-Stark analyses are complete. Chloroform extraction can be done in batches. Change the chloroform after the first 48 hours, and then change it every 24 hours thereafter. Continue chloroform extraction until clean chloroform shows no crude-oil color after 8 hours. Extraction time should be a minimum of 2.5 days for oil-zone cores. Cores from gas zones may require less time.

Note: If connate brine has a high salinity, the remaining salts can bias permeability, porosity, and other routine measurements. For samples not analyzed for tracer or brine composition, methanol extraction is recommended to remove salt crystals from the core samples to improve porosity, permeability, and grain-density accuracy.

Dry the core plug and thimble in a convection oven at 220°F at 1 atm for 4 hours. Then, dry the core plug and thimble for 24 hours at 220°F in a vacuum oven with a dry ice/acetone trap between the vacuum pump and oven. Check the weight and continue drying in 4-hour increments until a constant weight (±0.01 g) is obtained. If fragile clays like fibrous illite are present, dry the samples at 60°F in a vacuum oven to avoid breaking the fibers and artificially increasing the permeability of the rock. *Note: Damage may already have occurred during the Dean-Stark toluene extraction.* Cool the core plug and thimble over desiccant in a vacuum desiccator. Weigh the core plug and thimble to ±0.001 g. Remove the core plug from the thimble. Brush the core plug to remove fines, and then weigh the core plug again. Store the core plugs in individual containers over desiccant when not in the test equipment for porosity and/or permeability measurements. If the core plug weight increases by as much as 0.05 g from its original dry weight, then it must be dried again.

8.5.4 Porosity and Air Permeability. Before PV and permeability measurements, the grain volumes should be measured using a helium porosimeter calibrated daily using steel plugs of known volume. Three types of check samples (standard Berea sandstone, steel plugs, and lead or nickel foil) should be run with each set of test samples. The instrument should be recalibrated if any of these standards do not check satisfactorily. The grain volume data are used along with the PV measurement to calculate porosity at reservoir stress.

PV is usually measured at several stress (pressure) levels up to the mean reservoir net-overburden stress. Net stress is the overburden stress minus the supporting pore pressure. Measuring porosity and permeability at confining stress results in more accurate in-situ reservoir values. Confining pressure should always start at the lowest value and proceed to the highest value. The maximum-hydrostatic-overburden pressure is generally considered to be the mean net effective stress estimated from the net-overburden stress (vertical) multiplied by 0.67 to convert it into an approximate mean stress (i.e., assumes a Poisson's ratio of 0.33). If the Poisson's ratio is known, the mean net-reservoir stress can be calculated directly, instead of being estimated. The core holder applies stress equally in all directions (hydrostatic) while the rock in the reservoir has a vertical stress equal to the net overburden and two horizontal stresses typically less than the net overburden; therefore, the mean net-reservoir stress should be applied to the hydrostatic core holder. Plug bulk volume is calculated from the sum of the gas expansion PV and the grain volume measured in a helium pycnometer.

8.5.5 Mean Stress for Overburden Measurements. Hassler sleeve core holders operate at hydrostatic overburden conditions while the reservoir is modeled using three different orthogonal stresses—a vertical and two horizontal stresses. For making calculations, it is assumed that all total stresses (σ) have been converted into effective stresses (σ') or total stress minus pore pressure (P_ϕ) as follows:

$$\sigma' = \sigma - P_\phi \quad \text{(8.1)}$$

The vertical stress corresponds to the true vertical depth of the sample times the density gradient in psi/ft or psi/m of the overburden rock (rule of thumb: 1 psi/ft). For simplicity, assume the two horizontal stresses are equal as follows:

$$\sigma_{H1}' = \sigma_{H2}' = \sigma_H'. \quad \text{(8.2)}$$

In the field of rock mechanics, a constant called the stress path (K) is defined from the Poisson's ratio (v) of the rock as follows:

$$K = v / (1 - v). \quad \text{(8.3)}$$

Using the conventional elastic assumption, use this stress path times the vertical stress (σ_V') to define the horizontal stress (σ_H') as follows:

$$\sigma_H' = K \times \sigma_V'. \quad \text{(8.4)}$$

To approximate the reservoir stress conditions with a Hassler sleeve core holder, operate at mean stress (σ_M'), which equals the following:

$$\sigma_M' = 1/3\,(\sigma_V' + \sigma_H' + \sigma_H') = 1/3\,(\sigma_V' + 2\,\sigma_H'). \quad \text{(8.5)}$$

Using horizontal stress as a function of the stress path/Poisson's ratio, the mean stress equals the following:

$$\sigma_M' = 1/3\,(1 + 2K) \times \sigma_v' = 1/3\,[(1 + v) / (1 - v)] \times \sigma_V'. \quad \text{(8.6)}$$

Overburden conditions are simulated in Hassler sleeve core holders using mean stress instead of vertical stress. If the Poisson's ratio is not known, assume a Poisson's ratio of 0.33. This ratio calculates into a multiplier of 0.67 for mean stress calculations, which is reasonable for many rocks.

Automated PV and permeability measurements are preferred because the duplication of procedures is more precise and the results are more reproducible. It must be noted that exposure of core plugs to humidified laboratory air, which normally occurs with this equipment, must be minimized by purging the plastic enclosure around the carousel that holds the core plugs with dry nitrogen gas. Core plugs exposed to standard laboratory air for more than 30 minutes should be redried and reweighed. Dry core plugs should be stored in a desiccator containing fresh desiccant until the porosity and permeability measurements can be made. The equipment should be calibrated daily and also before any series of measurements. Every tray of samples (approximately 10) should be run with both a porosity and a permeability check plug. (Some laboratories run check plugs as often as every five samples.) The porosity check plug should be the first sample in line, so if the correct value (±0.1%) is not obtained, the instrument can be recalibrated.

Permeability to air is usually measured on all routine and vertical core plugs. Before this measurement is made, the fine materials generated while cutting the core plug should be cleaned from the plug faces with a brush.

8.5.6 Bulk Volume. Bulk volume is the sum of the grain volume determined with the helium pycnometer and the PV.

8.5.7 Tracer Analysis. Tracer-analysis procedures, calculations, and laboratory-analysis options are covered in Section 5.4, Appendix E, Appendix F, and Appendix J.

8.5.8 Core Photography. Gently clean the face of the ¼-core slab, and then photograph it in both white light and UV light. Depth marking should be readable on the core or expanded polystyrene packaging insert. Other markings should include operator, well name, scale bar, and color comparator. Photographing the core—both wet and dry—may assist in identifying some geological features. The sedimentologist describing the core may request close-ups of specific geological features. Repackage the ¼-core slab, and either forward it for geologic description, return it to its host country, or send it to the designated core-storage facility.

8.5.9 Destination for Data and Core Material. The core-analysis contractor should be given detailed instructions concerning distribution of the core-analysis data. These instructions usually include several copies for the operator and at least one copy for each partner in the well. Sometimes, the coring project manager acts as a central gathering point for the data until all the laboratory analyses are completed.

The remaining core material (the ¼-core slabs and preserved ¾-core butt carcasses) should be sent to the designated long-term storage facility. For cores shipped from one country to another, the ¼-core slabs are frequently returned to the country of origin for their permanent storage. The preserved ¾-core butt material can be the future source of rock for formation-damage tests, rock mechanics measurements, and other specialty rock analyses. This material is usually archived at the operator's core storage facility.

8.5.10 Chloride (Formation Brine) Analysis. The procedure for extracting and analyzing chlorides is given in Appendix H.

8.6 Poorly Consolidated Cores

Poorly consolidated cores must be held in their original configuration until sampling is completed. This stabilization can be accomplished by injecting materials such as epoxy or plaster of Paris into the annulus between the inner core barrel and the core. However, the preferred approach is freezing the core with dry ice. Chapter 7 describes core stabilization in more detail, but for the following paragraphs, the core is assumed to be frozen. *Note: For poorly consolidated cores, freezing boxes, insulated chests, dry ice, insulated gloves, and related materials must be added to the rigsite equipment lists.*

8.6.1 Core Analysis—Frozen Core. When the core is expected to be unconsolidated, full-closure-catcher core barrels must be used. At the rigsite, the core is frozen, preferably in 30-ft lengths using dry ice for at least 8 hours. The core-analysis laboratory is expected to supply equipment and personnel to the rigsite for the freezing of such cores. The frozen core, preserved inside the aluminum liner, is then cut into 3-ft lengths, capped and clamped, and freeze-stabilized in dry-ice chests for transport to the core-analysis laboratory. All core processing through the cutting of the core plugs must be performed with frozen-core techniques.

To perform total and spectral gamma ray analysis, the samples are left in the aluminum barrels (either the gamma ray detector must have high sensitivity or the analysis time must be appropriately increased). The samples should be at a temperature of –40°F or colder before placing them on the belt. The cores should be taken out of the freeze box, placed on the belt, analyzed, and immediately placed back in the freeze box. Every step must ensure that the cores start frozen and stay frozen while being handled.

The cores must be CT scanned to evaluate damaged zones, recovery, and bedding angle. Invasion of drilling fluids may also be determined. Two longitudinal scout views (i.e., standard and 90° to standard) should be taken from each 3-ft core segment. Five cross-sectional slices 2 mm thick should also be obtained from each 3-ft tube. These locations should be picked after reviewing the scout views. The core markings should be placed up for future orientations to be determined.

The frozen inner core barrel is split longitudinally into two pieces using a sawing jig and carbide-tipped saw set to penetrate the 0.25-in. aluminum wall. Make two saw cuts 180° apart for one-half the inner core barrel to be removed. The core should be kept frozen during this process. Use a polyvinyl chloride (PVC) tube (split in half to hold the core in place) in the aluminum half-tube, and slab the core ¼ to ¾ using liquid nitrogen to keep the core frozen while slabbing with a band saw. The bedding angle may be observable on the exposed ends of the core (or from the CT-scan data). If possible, the core should be slabbed ¼ to ¾ for the bedding and updip angle to be clearly observable. Place the ¼-core slab in a slab box with expanded polystyrene packaging inserts. The ¼-core slab is not kept at dry-ice temperature but should be kept in a freezer at –20°F and kept covered until it can be photographed. The ¾-core slab must be kept at dry-ice temperature until the operator point contact decides it can be placed in a freezer. The ¾-slab remains cradled in the inner-barrel half, is spiral wrapped in plastic wrap, and is placed in PVC tubes, capped, and clamped when placed on dry ice for refreezing. Mark depths on the face of the ¾-core slab. The inner-barrel half should retain depth and orientation marks. Mark depths and orientation on the PVC tube. *Note: Be aware that extended time at these low temperatures can result in significant sublimation.* Spraying 3% KCl brine on the slabbed surface before wrapping with plastic wrap creates an ice cake to protect the rock from sublimation damage.

Some laboratories have had success with slabbing in one pass with a band-saw blade designed to cut both aluminum and rock.

Using liquid nitrogen as the plugging fluid, the 1.5-in. diameter routine core plugs should be cut at each foot from the ¾-core butt core center. A milling machine is preferred because it produces less vibration than a drill press. These core plugs are immediately trimmed to 1.5-in. length. Minimize exposure of the core plugs to laboratory air; moisture can condense on them because of their very cold temperature, causing errors with the water saturation being too high. Each trim end (i.e., the outer part of the core) is wrapped in plastic wrap, placed in a glass container, capped to seal, and kept in the freezer at –20°F until distribution for either thin section, XRD, or other testing. The trimmed core plugs are wrapped in thread seal tape and encased in a nickel foil sleeve with two screens (one fine and one coarse) on each end. Weights must be taken at each step of the encasing process. The core plugs may then be placed into a hydrostatic core holder (at the mean effective confining stress level selected by the operator point contact) and then thawed. *Note: Some operators prefer to initially seat the foil sleeve with the sample frozen (approximately 400 psi), allowing the thawing to occur unstressed to avoid possible rearrangement caused by uneven thawing.*

If fluids are produced while thawing the core plugs, these fluids must be collected and weighed. Note and record whether these fluids are oil or water. Obtain final core plug weights. The core plugs can then be preserved in the same manner as the routine core plugs from consolidated rock. The remainder of the procedures do not differ significantly from those for consolidated rock intervals, except when grain densities are measured.

The SCAL core-plug intervals should be selected from the most homogeneous rock sections. Mechanical properties are one of the more important tests to which SCAL plugs are allocated in unconsolidated intervals. These SCAL plugs should be vertical and 1.5 or 1.0 in. in diameter by 3.0 or 2.0 in. long to provide the proper length-to-diameter ratio for such tests. To obtain vertical samples, saw a piece of core of the appropriate length (4.0 in. for mechanical tests), and then cut the samples. Multiple vertical samples are typically taken. Other SCAL core plugs for flow and resistivity measurements should be horizontal and 1.5 in. in diameter by 2.0 in. long. Save trim ends, and preserve the core plugs as previously discussed in Section 8.4. For the mechanical-property core plugs, the net confining stress during stabilization should not exceed 400 psi because too high a compaction stress may affect the test results.

After all measurements are completed on unconsolidated-sand RCAL core plugs, the nickel sleeves and screens can be removed from the samples. (If the samples are needed for future tests, grain volumes can be obtained by helium porosimetry in which the volumes of metal, tape, and screen are subtracted.) They should then be dried to a constant weight in a vacuum oven at 220°F, cooled in a vacuum oven, and then the grain densities should be remeasured on the unmounted samples. This grain density on each core plug should be the reported value; if the original grain density calculated for the mounted sample is different by less than 0.01 g/cm^3, then use this grain density to calculate a corrected porosity value. The grain volume should be based on the post-Dean-Stark-water-extraction dry-sample weights.

The disaggregated, crushed, and leached grains from each core plug from the tracer analysis are saved in a labeled and capped bottle for possible later use in grain-size analysis. Selected samples may be used to cover the range of rock types expected. The grains from selected core plugs are collected on 0.45-μm analytical-grade filter paper, washed with distilled water to remove salt, and dried and disaggregated for grain-size analysis. The operator point contact for RCAL is responsible for the selection of the core plugs to be analyzed.

It may be necessary to clean the core face of saw debris after slabbing. Place the ¼-core slab in a slab pack. Take color photographs of the ¼-core slab in white and UV light. The photograph has a higher quality if the surface of the core does not have ice on it; therefore, the stabilized ¼-core slab should be photographed immediately or after it is fully thawed. Determine which status provides the best pictures. When taking the photographs, proper depth markings should be readable on the core or expanded polystyrene packaging insert.

8.7 Contacts

When beginning the planning of an LIC project, the coring project manager and the operator's core-analysis specialist should establish a good working relationship with key contacts at the core-analysis laboratory. These people are very important in the detailed planning of the operation. Frequent and detailed communications should begin as soon as the contractor analysis laboratory is chosen.

Nomenclature

k = permeability, md
K = stress path, dimensionless (Eq. 8.3)
N_c = capillary number, dimensionless
P_ϕ = pore or reservoir pressure, psi (Eq. 8.1)
R_w = water resistivity, Ω·m
S_{cw} = connate water saturation, %PV
S_o = oil saturation, %PV
S_{or} = residual oil saturation, %PV
S_{org} = residual oil saturation to gasflooding, %PV
S_{orw} = residual oil saturation to water displacement, %PV
S_w = water saturation, %PV
σ = total stress, psi (Eq. 8.1)
σ' = effective stress, psi (Eq. 8.1)
σ_H = horizontal stress, psi (Eq. 8.2)
σ_{H1} = principal horizontal stress, psi (Eq. 8.2)
σ_{H2} = horizontal stress orthogonal to principal horizontal stress, psi (Eq. 8.2)
σ_M = mean stress, psi (defined in Eq. 8.5)
σ_V' = vertical stress, psi (Eqs. 8.4 and 8.5)
ν = Poisson's ratio, dimensionless (Eq. 8.3)
ϕ = subscript referring to reservoir conditions or within pore space (Eq. 8.1)

Glossary

API	=	American Petroleum Institute
BHA	=	bottomhole assembly
BHI	=	Baker Hughes Inteq
CT	=	computer tomography
DAFD	=	downhole-activated flow diverter
DEA	=	Drilling Engineering Association
DOT	=	Department of Transportation
ECD	=	electron-capture detector
EOR	=	enhanced oil recovery
EP	=	end piece
ES	=	electrical stability
FID	=	flame-ionization detector
GC	=	gas chromatograph
HP/HT	=	high pressure/high temperature
HTO	=	water with tritium replacing one hydrogen atom
ID	=	internal diameter, in.
IFT	=	interfacial tension, dynes/cm
JIP	=	joint industry project
LCM	=	lost-circulation material
LD	=	long distance
LIC	=	low-invasion coring
LSRV	=	low-shear-rate viscosity
M	=	modified
MD	=	measured depth
MWD	=	measurement while drilling
OBM	=	oil-based mud
OD	=	outside diameter, in.
OWC	=	oil/water contact
PAC	=	polyanionic cellulose
PDC	=	polycrystalline diamond compact
PV	=	pore volume, cm^3
PVT	=	pressure/volume/temperature
QC/QA	=	quality control/quality assurance
RCAL	=	routine-core-analysis laboratory
REP	=	routine core plug end piece
RIH	=	run in hole
ROP	=	rate of penetration, ft/hour
RPM	=	rotations per minute
SCAL	=	special-core-analysis laboratory
TFA	=	total flow area, $in.^2$
TVDSS	=	true vertical depth subsea
USD	=	United States dollar
UV	=	ultraviolet

WBM	=	water-based mud
WOB	=	weight on bit
XO	=	crossover
XRD	=	X-ray diffraction
YP	=	yield point
μL	=	microliter

References

Abrams, A. 1975. The Influence of Fluid Viscosity, Interfacial Tension, and Flow Velocity on Residual Oil Saturation Left by Waterflood. *SPE J* **15** (5): 437–447. SPE-5050-PA. DOI: 10.2118/5050-PA.

American Petroleum Institute. *API RP-40: Recommended Practices for Core Analysis,* Second Edition, February 1998.

Ballard, T.J. and Dawe, R.A. 1988. Wettability Alteration Induced by Oil-Based Drilling Fluid. Paper SPE 17160 presented at SPE Formation Damage Control Symposium, Bakersfield, 8–9 February. DOI: 10.2118/17160-MS.

Bloys, J.B., Rathmell, J.J., Kriel, W.A., Spence, A.P., Wylie, G.S., and Ray, C.R. 1993. Development and Application of a Mineral-Oil-Based Mud and Tracer for Low-Invasion Coring at the Point McIntyre Field. Paper SPE 26326 presented at the SPE Annual Technical Conference and Exhibition, Houston, 3–6 October. DOI: 10.2118/26326-MS.

Charlson, G.S., DeRuiter, R.A., Spence, A.P., and Warner, H.R. Jr. 1997. Application of Oil-Base-Mud Pressure Coring to the Determination of In-Situ Water Saturations. *SPE Drill & Compl* **12** (2): 105–110. SPE-28595-PA. DOI: 10.2118/28595-PA.

Clydesdale, G.M., Leseultre, A., and Lamine, E. 1994. Low Invasion Corehead Reduces Mud Invasion While Improving Performances. *Journal of Energy Resources Technology* **116:** 258–267.

Dawe, B.A. and Murdock, D.M. 1990. Laminated Sands: An Assessment of Log Interpretation Accuracy by an Oil-Base Mud Coring Program. Paper SPE 20542 presented at the SPE Annual Technical Conference and Exhibition, New Orleans, 23–26 September. DOI: 10.2118/20542-MS.

Dombrowski, H.S. and Brownell, L.E. 1954. Residual Equilibrium Saturation of Porous Media. *Industrial and Engineering Chemistry* (June 1954): 1207–1219.

Donaldson, E.C. and Siddiqui, T.K. 1989. Relationship Between the Archie Saturation Exponent and Wettability. *SPE Form Eval* **4** (3): 359–362. SPE-16790-PA. DOI: 10.2118/16790-PA.

Egbogah, E.O. and Amar, Z.H. Bin Tajul. 1997. Accurate Initial/Residual Saturation Determination Reduces Uncertainty in Further Development and Reservoir Management of the Dulang Field, Offshore Peninsular Malaysia. Paper SPE 38024 presented at the SPE Asia Pacific Oil and Gas Conference and Exhibition, Kuala Lumpur, 14–16 April. DOI: 10.2118/38024-MS.

Holstein, E.D. and Warner, H.R. Jr. 1994. Overview of Water Saturation Determination for the Ivishak (Sadlerochit) Reservoir, Prudhoe Bay Field. Paper SPE 28573 presented at the SPE Annual Technical Conference and Exhibition, New Orleans, 25–28 September. DOI: 10.2118/28573-MS.

Huff, C.F., McFall, A.L., and St. Clair, J.A. 1978. Design of Special Performance Bits Utilizing Synthetic Diamond Cutters. Paper SPE 7515 presented at the SPE Annual Technical Conference and Exhibition, Houston, 1–3 October. DOI: 10.2118/7515-MS.

Jenks, L.H., Huppler, J.D., Morrow, N.R., and Salathiel, R.A. 1968. Fluid Flow within a Porous Medium Near a Diamond Core Bit. *J Can Pet Technol* (October–December): 172–180.

McCoy, D.D., Warner, H.R. Jr., and Fisher, T.E. 1997. Water-Salinity Variations in the Ivishak and Sag River Reservoirs at Prudhoe Bay. *SPE Res Eng* **12** (1): 37–44. SPE-28577-PA. DOI: 10.2118/28577-PA.

Mungan, N. and Moore, E.J. 1968. Certain Wettability Effects on Electrical Resistivity in Porous Media. *J Can Pet Technol* (January–March): 20–25.

O'Sullivan, T., Zittel, R.J., Beliveau, D., Wheaton, S., Warner, H.R. Jr., Woodhouse, R., and Ananthkrishnan, B. 2008. Very Low Water Saturations Within the Sandstones of the Northern Barmer Basin, India. Paper SPE 113162 presented at the SPE Indian Oil and Gas Technical Conference and Exhibition, Mumbai, India, 4–6 March. DOI: 10.2118/113162-MS.

Pallatt, N., Stockden, I., Mitchell, P., and Woodhouse, R. 1991. Low Invasion Coring Gives "Native" Reservoir Water Saturations. In *Proc.,* 14th European Formation Evaluation Symposium, London, 1–16.

Rathmell, J.J., Atkins, L.K., and Kralik, J.G. 1999. Application of Low Invasion Coring and Outcrop Studies to Reservoir Development Planning for the Villano Field. Paper SPE 53718 presented at the SPE Latin American and Caribbean Petroleum Engineering Conference, Caracas, 21–23 April. DOI: 10.2118/53718-MS.

Rathmell, J.J., Bloys, J.B., Bulling, T.P., Parr, J.J., Pasternack, E.S., Jerauld, G.R., and Sakurai, S. 1995. Low Invasion, Synthetic Oil-Base Mud Coring in the Yacheng 13-1 Gas Reservoir for Gas-in-Place Calculation. Paper SPE 29985 presented at the SPE International Meeting on Petroleum Engineering, Beijing, 14–17 November. DOI: 10.2118/29985-MS.

Rathmell, J.J., Gremley, R.R., and Tibbitts, G.A. 1994. Field Applications of Low Invasion Coring. Paper SPE 27045 presented at the SPE Latin American and Caribbean Petroleum Engineering Conference, Buenos Aires, 27–29 April. DOI: 10.2118/27045-MS.

Rathmell, J.J., Tibbitts, G.A., Gremley, R.B., Warner, H.R. Jr., and White, E.K. 1995. Development of a Method for Partially Uninvaded Coring in High-Permeability Sandstone. *SPE Form Eval* **10** (2): 65–71. SPE-20413-PA. DOI: 10.2118/20413-PA.

Rathmell, J.J., Wilton, B.S., Gale, B.A., Bell, D.A., Tibbitts, G.A., and Bobrosky, D.J. 1998. Development and Application of PDC Core Bits for Down-hole Low Invasion Coring in the Arab Carbonates. *SPE Drill & Compl* **13** (1): 56–65. SPE-36263-PA. DOI: 10.2118/36263-PA.

Reed, R.L., Humphrey, R.A., Holstein, E.D., and Rathmell, J.J. 1994. IFT and Flushing Studies To Validate As-Received Oil-Base-Core Water Saturations From the Ivishak (Sadlerochit) Reservoir, Prudhoe Bay Field. Paper SPE 28594 presented at the SPE Annual Technical Conference and Exhibition, New Orleans, 25–28 September. DOI: 10.2118/28594-MS.

Richardson, J.G, Holstein, E.D., Rathmell, J.J. and Warner, H.R. Jr. 1997. Validation of As-Received Oil-Based-Core Water Saturations from Prudhoe Bay. *SPE Reservoir Engineering* **12** (1): 31–36. DOI: 10.2118/28592-PA.

Sanner, D.O. and Azar, J.J. 1994. Alteration of Reservoir Rock Wettability and its Flow Properties Caused by Oil-Based and Water-Based Drilling Muds. Paper SPE 27354 presented at SPE Formation Damage Control Symposium, Lafayette, 7–10 February. DOI: 10.2118/27354-MS.

Sharma, M.M. and Wunderlich, R.W. 1985. The Alternation of Rock Properties due to Interaction with Drilling Fluid Components. Paper SPE 14302 presented at the SPE Annual Technical Conference and Exhibition, Las Vegas, Nevada, USA, 22–26 September. DOI:10.2118/14302-MS.

Stalkup, F.I. 1983. *Miscible Displacement.* Monograph Series, SPE, Richardson, Texas **8.**

Thomas, D.C., Hsing, H. and Menzie, D.E. 1984. Evaluation of Core Damage Caused by Oil-Based Drilling and Coring Fluids. Paper SPE 13097 presented at the SPE Annual Technical Conference and Exhibition, Houston, 16–19 September. DOI: 10.2118/13097-MS.

Tibbitts, G.A. and Radford, S.R. 1985. New Technology and Tools for Recovery of Representative Cores From Uncemented Sandstone Formations. Paper SPE 14297 presented at the SPE Annual Technical Conference and Exhibition, Las Vegas, Nevada, USA, 22–26 September. DOI: 10.2118/14297-MS.

Tibbitts, G.A., Reed, M.G., and McCarter, M. 1990. New Coring System Reduces Filtrate Invasion. *Advances in Core Evaluation, Accuracy and Precision in Reserves Estimation,* Society of Core Analysts First European Core Analysis Proceedings, 55–77. Philadelphia, Pennsylvania, USA: Gordon and Breach Science Publishers.

Walker, B.H. and Black, A.D. 1993. Dynamic Spurt-Loss Beneath an Oilfield Bit. Paper SPE 25700 presented at the SPE/IADC Drilling Conference, Amsterdam, 22–25 February. DOI: 10.2118/25700-MS.

Warner, H.R. Jr. and Rathmell, J.J. 1997. Mechanisms Controlling Filtration at the Core Bit for Oil-Based Muds. *SPE Drill & Compl* **12** (2): 111–118. SPE-28596-PA. DOI: 10.2118/28596-PA.

Woodhouse, R. 1998. Accurate Reservoir Water Saturations From Oil-Mud Cores: Questions and Answers From Prudhoe Bay and Beyond. *The Log Analyst* (May–June): 23–47.

Wylie, G.S., Rathmell, J.J., and Tibbitts, G.A. 1993. Coring Technology for Point McIntyre Wells. Paper SPE 26353 presented at the SPE Annual Technical Conference and Exhibition, Houston, 3–6 October. DOI: 10.2118/26353-MS.

Yan, J-N., Monezes, J.L. and Sharma, M.M. 1993. Wettability Alteration Caused by Oil-Based Muds and Mud Components. *SPE Drill & Compl* **8** (1): 35–44. SPE-18162-PA. DOI: 10.2118/18162-PA.

Appendix A

Selected SPE Papers Concerning LIC Technologies

Note: *Complete versions of these papers can be found at www.onepetro.org.*

Tibbitts, G.A. and Radford, S.A. 1985. New Technology and Tools for the Recovery of Representative Cores from Uncemented Sand Formations. Paper SPE 14297 presented at the SPE Annual Technical Conference and Exhibition, Las Vegas, Nevada, USA, 22–26 September. DOI: 10.2118/14297-MS.

- Study describing tests run with the goal of achieving better core recovery of unconsolidated Faja sand in the Hamaca area of Venezuela—25 full-scale coring tests.
- Sand fluidized when standard coring parameters used.
- Use of face discharge, PDC bit.
- Use of limited ROP to allow filter cake time to form.
- Use of low-spurt-loss coring mud.
- Use of low mud flow rate—minimum to clean bit.
- Use of extended-catcher shoe.
- Use of full-closure core barrel.

Rathmell, J.J., Tibbitts, G.A., Gremley, R.B., Warner, H.R. Jr., and White, E.K. 1995. Development of a Method for Partially Uninvaded Coring in High-Permeability Sandstone. *SPE Form Eval* **10** (2): 65–71. DOI: 10.2118/20413-PA.

- First LIC achieved using WBM. Cut 4-in. cores with 2.5-in. uninvaded centers. Previous technical papers described low filtrate invasion only being achieved with OBM.
- WBM contained significant concentrations of fine calcium carbonate for rapid filter-cake formation.
- New style PDC core bit (first ARC-412) with face-discharge ports, extended catcher shoe, no secondary throat gauge cutters, parabolic profile, and 1/2-in. cutters.
- ROP > 90 ft/hr. Low invasion found at ROP as low as 30 ft/hr.
- Restored-state cores probably not adequate for accurate relative permeability measurements. LIC needed for native-state measurements.

Walker, B.H. and Black, A.D. 1993. Dynamic Spurt-Loss Beneath an Oilfield Bit. Paper SPE 25700 presented at the SPE/IADC Drilling Conference, Amsterdam, 22–25 February. DOI: 10.2118/25700-MS.

- Summary of cooperative industry projects DEA-23 and DEA-57; spurt-loss tests run at TerraTek facilities in Salt Lake City, Utah, USA.
- Details of the equipment and procedures used.

- Highlights the need for high ROP.
- Does not include mud design conclusions.

Bloys, J.B., Rathmell, J.J., Kriel, W.A., Spence, A.P., Wylie, G.S., and Ray, C.R. 1993. Development and Application of a Mineral-Oil-Based Mud and Tracer for Low-Invasion Coring at the Point McIntyre Field. Paper SPE 26326 presented at the SPE Annual Technical Conference and Exhibition, Houston, 3–6 October. DOI: 10.2118/26326-MS.

- Low-invasion OBM formulation.
- Iodonaphthalene as tracer (bromonaphthalene now preferred).
- Spacers/displacement from WBM to OBM coring fluid.
- Coring fluid maintenance insights.
- Example of resolving equity redetermination more quickly with accurate oil-in-place determination.

Wylie, G.S., Rathmell, J.J., and Tibbitts, G.A. 1993. Coring Technology for Point McIntyre Wells. Paper SPE 26353 presented at the SPE Annual Technical Conference and Exhibition, Houston, 3–6 October. DOI: 10.2118/26353-MS.

- Coring-bit design for harder rock—full-scale pretesting.
- Total of 3,068 ft cored in eight wells—99.6% core recovery, < 2% of core plugs invaded.
- Arctic rigsite.
- Core processing on site.
- Core-analysis procedures.
- Connate water cannot be considered immobile if saturation > 65%.
- Data was of such high quality that five wells were eliminated from the overall coring program to finalize equity percentages.

Reed, R.L., Humphrey, R.A., Holstein, E.D., and Rathmell, J.J. 1994. IFT and Flushing Studies to Validate As-Received Oil-Base-Core Water Saturations from the Ivishak (Sadlerochit) Reservoir, Prudhoe Bay Field. Paper SPE 28594 presented at the SPE Annual Technical Conference and Exhibition, New Orleans, 25–28 September. DOI: 10.2118/28594-MS.

- Laboratory tests to determine the IFT values for various mud-filtrate/brine combinations.
- Laboratory flow tests to determine the conditions under which connate water was mobilized.
- Concluded IFT values were dependent on contact times, and higher IFT found in short-duration tests mimicking the time taken to actually cut a core.
- All tests at realistic reservoir conditions resulted in insignificant water mobilized.
- Tests indicated both high-pressure gradients and high-carboxylate concentrations were needed to mobilize connate water.

Warner, H.R. Jr. and Rathmell, J.J. 1997. Mechanisms Controlling Filtration at the Core Bit for Oil-Based Muds. *SPE Drill & Compl* **12** (2): 111–118. DOI: 10.2118/28596-PA.

- Half-scale laboratory tests—4-in. cores used in minicoring tests to cut 2-in. minicores.
- A good OBM filter cake takes up almost all the differential pressure experienced by the core.
- OBM filtrate loss rates low in oil-saturated cores—unlikely to move connate water.
- Filtrate/formation IFT (interfacial tension) brine only approximately half of reservoir crude-oil/brine IFT.
- Above the OWC and oil/water-transition zone, connate water not likely to be mobilized by OBM filtrate.

Rathmell, J.J., Bloys, J.B., Bulling, T.P., Parr, J.J., Pasternack, E.S., Jerauld, G.R., and Sakurai, S. 1995. Low Invasion, Synthetic Oil-Base Mud Coring in the Yacheng 13-1 Gas Reservoir for Gas-in-Place Calculation. Paper SPE 29985 presented at the SPE International Meeting on Petroleum Engineering, Beijing, 14–17 November. DOI: 10.2118/29985-MS.

- Remote operations—South China Sea offshore drilling platform.
- Coring procedures—rig core-processing procedures.
- High-temperature (350°F) environment with OBM.
- Low-filtrate-invasion core identified a salinity variation within the reservoir interval—needed for more accurate log calibrations and calculations.
- 1,100-psi overbalance.
- Very low formation damage.
- Mostly used AXC-412 coring bits.
- Countercurrent imbibition increases apparent OBM filtrate invasion in gas-reservoir interval, but should not mobilize connate water.
- Laboratory core-processing procedures.

Rathmell, J.J., Wilton, B.S., Gale, B.A., Bell, D.A., Tibbitts, G.A., and Bobrosky, D.J. 1998. Development and Application of PDC Core Bits for Downhole Low Invasion Coring in the Arab Carbonates. *SPE Drill & Compl* **13** (1): 56–65. DOI: 10.2118/36263-PA.

- Water-based LIC fluid formulation with iodide ion tracer.
- Bacteria problems—need for higher biocide concentrations (0.5-ppb glutaraldehyde).
- Coring with mud motor to double ROP in carbonate formations.
- Use of circulating sub to circulate bottoms up at end of coring run without rotating mud motor and core barrel.
- Remote operations—offshore Qatar.
- Very low formation damage.
- Antiwhirl bit features were critical at high rev/min.
- Modified, more rounded profile AXC-412 coring bit required for harder rock, with concentrically reinforced cutters—20° back-rake on core-bit face cutters.
- H_2S provisions.
- Upgrade required of local core-analysis laboratory.

Rathmell, J.J., Atkins, L.K., and Kralik, J.G. 1999. Application of Low Invasion Coring and Outcrop Studies to Reservoir Development Planning for the Villano Field. Paper SPE 53718 presented at the SPE Latin American and Caribbean Petroleum Engineering Conference, Caracas, 21–23 April. DOI: 10.2118/53718-MS.

- Remote operations (Ecuadorian rain forest, helicopter access only).
- Previous cores showed unrealistically high water saturations when cored with standard WBM.
- 86% of plugs had low filtrate invasion—accurate water saturations.
- Water-based field mud adapted for LIC—bromide tracer.
- Mud pressure overbalanced 1,800 psi.
- Permeability streaks up to 3 darcies—fine calcium carbonate still performed satisfactorily to develop acceptable filter cake.
- Salinity gradient detected—2,600-ppm chlorides at base of reservoir interval, with 36,000-ppm chlorides at top—led to improved log calibration and calculations.
- 428 ft of core cut—100% recovery—45 ft/hr.
- Data key for optimizing field development plan.

Skopec, R.A., Collee, P.E., van Puymbroeck, L., and Shallenberger, L.K. 1995. Downhole Core Preservation for Improved Formation Evaluation. Paper presented at the SPWLA 36th Annual Logging Symposium, Paris, 26–29 June.

- Gel coring supplements other LIC technologies.
- Inner core barrel preloaded with high-viscosity polypropylene glycol gel.
- As core enters the inner core barrel, a floating piston is pushed back dispensing the gel and coating the core. The force necessary to push back the piston means gel coring cannot be done on weak or unconsolidated rock.
- The gel coating isolates the core from differential pressure and stops further filtrate invasion from static filtration, even in rock with permeabilities up to 10 darcies.
- Especially helps when coring ROP is low, or when differential pressures are high. Both cases otherwise lead to higher filtrate invasion.
- Does not prevent initial filtrate invasion during the spurt loss as core is first cut and filter cake is being formed.
- A North Sea field example is given where improved water saturation accuracy was obtained.

Appendix B

Special Circumstances—Sponge Coring and the Use of Noninvading Core Gel

Two special circumstances may occasionally arise in LIC projects.

B.1 Gel Coring

The first special circumstance involves using viscous polypropylene glycol gel to automatically coat cores after they enter the inner core barrel of the coring assembly. This coating minimizes the static invasion of mud filtrate into the core while the core is still experiencing mud overbalance pressure. Note that the gel coats the core only after it is a foot or more inside the inner core barrel, so the mud filtrate that enters the core with the initial spurt loss is not affected. The system is particularly useful for significantly slower coring rates common in hard rock, when the core will experience differential mud pressure for a longer time period or when the differential mud pressure is very high, such as coring a pressure-depleted reservoir. As a bonus, the gel is also a good lubricant and can substantially reduce jamming when coring fractured rock. The Gel Coring™ system is a Baker Hughes product. The gel is preloaded into special 30-ft sections of inner core barrel and is slowly dispensed as the top of a core pushes back a floating piston. Because of the force needed to push back the piston, gel coring is not recommended for soft or unconsolidated formations. The gel must be removed from the core in the core-analysis laboratory before slabbing and plugging can begin. It may be necessary to scrape the bulk of the gel from the core using a ¼-cylindrical section of PVC pipe having a diameter similar to that of the core, before a final cleaning with rags. Core sampling and analysis do not change from the procedures described in Chapter 8.

B.2 Sponge Coring

As cores are pulled to the surface from downhole, there is a drop in hydrostatic pressure, which means that dissolved gases will be liberated and will expand, and then fluids (gas, oil, and/or water) will be expelled from of the core. Special-core inner-core barrels may be fitted with sponge liners to capture these fluids. The main reason for sponge coring is usually to try to obtain an accurate measurement of the volume of oil in the core. One example is coring to quantify the oil remaining after an enhanced-oil-recovery process has swept the reservoir where the core is cut. The sponge is usually oil-wet polyurethane, which readily absorbs and traps the oil phase. Where quantifying oil is the objective, water-based coring fluids are used to avoid contaminating the core and sponge with OBM filtrate. The sponges and hardware are supplied by the coring equipment contractor. The inner core barrel/sponge assemblies will arrive at the rigsite ready to use. Note that sponge cores are only 3.25 in. in diameter so that the inner core barrel/sponge assembly will fit in a normal outer core barrel.

Note that water-wet sponges can be used if collecting any mobile water phase is desired. In this case, oil-based coring fluids can be considered.

- **Rigsite Operations**
  - The inner core barrel is still cut into 3-ft sections. Sponge liners and core are preserved together in ProtecCore® laminate and placed in 6-in. PVC pipe with caps and clamps to seal the PVC. The tubes are then placed in boxes for transport to the contract core-analysis laboratory.

- **Laboratory Operations**

  Analyzing for total oil in the core requires that both the sponge and the adjacent core be analyzed. The core will be analyzed as outlined in Chapter 8, but a few additional instructions are noted here:
  - Conduct spectral gamma ray measurements while the core is still in the aluminum inner core barrel.
  - Split the aluminum inner core barrel with the slabbing saw, using a very shallow depth of cut to slice through the aluminum while avoiding cutting the sponge. Make two cuts 180° apart and remove the inner core barrel sections.
  - Most sponge core samples are analyzed as whole core, but if RCAL, SCAL, or tracer sample sections are desired, please refer to Chapter 8 for sampling and preservation instructions. Sponge sections that need to be preserved can be stored in ProtecCore® laminate pouches.
  - If RCAL samples are to be taken, mark these sections on the sponge liner.
  - Mark the sections for depth and core orientation.
  - Photograph any oil staining on both sides of the sponge.
  - Minimize oil evaporation by never exposing the sponge for longer than 30 minutes before representing.
  - Analysis of sponge liner:
    - Develop a calibration curve.
      - Using samples of produced crude oil from the same field, prepare solutions of the crude oil in hexane ranging from 0.1 to 2.0 cm^3 of crude oil per 100 g of reagent grade hexane (new, do not reclaim). Determine the adsorption spectrum on an ultraviolet (UV) spectrophotometer for each of the solutions. Select the wavelength having the greatest sensitivity to the crude-oil concentration; the wavelength should be at an adsorption peak rather than on a shoulder. This wavelength will be used to quantify the crude oil in the sponges. If the crude-oil composition is variable, then test the sensitivity of UV adsorption to the variation in composition or API gravity of the crude. It may be acceptable to use an average crude-oil gravity.
      - Once the adsorption peak has been selected, take samples of the sponge liner material being used in the field (same lot number) and separate the sponge from five or six 4-in.-long sections of aluminum. This will produce ten 1-in.-wide by 4-in.-long strips of sponge for each section. Prepare 10 strips of sponge liner to be extracted using a needle, and syringe amounts of crude oil from 1.0 cm^3 (0.1 cm^3 per strip of sponge) to 20 cm^3 (2 cm^3 per strip of sponge). Prepare a calibration curve consisting of at least five different volumes of crude oil within the specified range. Extract each group of 10 strips of sponge in a batch using 1,500 g of hexane for 1 hour, 2 hours, and 4 hours. Take a 1.0-cm^3 aliquot of the hexane after each of these times and measure its adsorption. Determine the time required for maximum oil extraction based on the time required to reach a constant value.
    - Analyze sponge from the core.
      - Remove the sponge liner that was opposite the routine whole-core sample from its preservation materials. Cut the sponge away from the aluminum liner. The sponge will be cut into strips 4 in. long and approximately 1 in. wide. The width of the strip is determined by the separation between the vanes in the aluminum liner.

The aluminum liner is not required for the analysis, and it may be discarded provided that the identification of the sponge material being analyzed is retained. Mark the sample identification on the 1-gal tin-coated paint can. Place the sponge material in the tared paint can, obtain the weight, add 1,500 g of reagent grade hexane (new solvent, not recycled), obtain weight, and shake on a paint mixer for the length of time determined previously. Mark the sample identification on the 1-gal tin-coated paint can. Take a 1.0-cm^3 aliquot and measure the UV absorbency at the predetermined wavelength. Resume agitation for an additional increment of time, and take a 1-cm^3 aliquot and measure the UV absorbency. If constant within 5%, discontinue extraction. After the first sample or two, the time required for extraction should have been confirmed, and intermediate sampling steps may be discontinued. Use the calibration curve to obtain the oil volume; this oil volume will be added to the oil in the whole-core sample by assuming it is lost equally from all portions of the original whole-core PV.

Appendix C

Calculation of Filtrate Saturation

C.1 Definition of Terms

C_m = concentration of tracer in the coring fluid; this value is supplied daily in digital format from the tracer-analysis laboratory, ppm
C_{tt} = concentration of tracer in the toluene from the Dean-Stark water extraction; this value is supplied daily in digital format from the tracer-analysis laboratory, ppm
F_f = weight fraction of filtrate in the oil in the core
S_{of} = saturation of filtrate in core, %PV
V_p = PV of the core, cm^3
W_{cd} = weight of dry core, g
W_{ci} = weight of core before Dean-Stark water extraction, g
W_{co} = weight of core after Dean-Stark water extraction, g
W_{f1} = weight of dry Dean-Stark flask, g
W_{loss} = weight of toluene lost as vapor, g
W_o = weight of oil phase in the core before Dean-Stark water extraction, g
W_{oe} = weight of crude oil extracted from core into toluene, g
W_{of} = weight of filtrate extracted from core, g
W_{op} = weight of plugging oil extracted from core, g
W_{or} = weight of residual oil left in core after Dean-Stark water extraction, g
W_{t1} = weight of flask and toluene before Dean-Stark water extraction, g
W_{t2} = weight of toluene, crude oil, filtrate, plugging oil, and flask after Dean-Stark water extraction, g
W_{tc} = weight of toluene remaining in core after Dean-Stark water extraction, g
W_{ti} = weight of toluene charged to flask before Dean-Stark water extraction, g
W_{to} = weight of toluene and other hydrocarbons remaining in the flask after Dean-Stark water extraction, g
W_w = weight of water distilled from core, g

The underlined terms are measured in the analysis. The other terms may be calculated, but not all these calculated terms are needed for the final report. Those terms needed are identified subsequently.

C.2 Analysis Procedure

Eq. C-1 calculates the weight of toluene charged to the Dean-Stark flask as follows:

$$W_{ti} = W_{t1} - W_{f1}. \quad \text{(C-1)}$$

Eq. C-2 calculates the weight of toluene plus other hydrocarbons in the flask after Dean-Stark water extraction is complete (includes the toluene in side arm) as follows:

$$W_{tc} = W_{t2} - W_{f1}. \qquad \text{(C-2)}$$

Eq. C-3 calculates the weight of toluene as the sum of all fluids in the flask after Dean-Stark water extraction as follows:

$$W_{tc} = W_{ti} + W_{oe} + W_{op} + W_{of} - W_{\text{loss}} - W_{tc}. \qquad \text{(C-3)}$$

Eq. C-4 calculates the total weight of oil phase in the core before Dean-Stark water extraction as follows:

$$W_o = W_{ci} - W_{cd} - W_w. \qquad \text{(C-4)}$$

Eq. C-5 calculates the total weight of the oil phase in the core before Dean-Stark water extraction as follows:

$$W_o = W_{oe} + W_{or} + W_{op} + W_{of}. \qquad \text{(C-5)}$$

Eq. C-6 calculates the toluene vapor lost from the Dean-Stark water extractor in terms of the other weights that are measured. This equation is the result of combining Eqs. C-1 through C-5 as follows:

$$W_{\text{loss}} = W_{t1} - W_{t2} + W_{ci} - W_{co} - W_w. \qquad \text{(C-6)}$$

Eq. C-7 calculates the mud-filtrate weight fraction in the oil in the core as follows:

$$F_f = (C_{tt} / C_m) \times (W_{to} / W_o). \qquad \text{(C-7)}$$

Eq. C-8 calculates the saturation of mud filtrate in the core. This value should be printed out with the routine core data as follows:

$$S_{of} = 100 \times [W_o \times (F_f / 0.82)] / V_p. \qquad \text{(C-8)}$$

The value 0.82 g/cm^3 is the estimated mud-filtrate density. There are no significant assumptions in the calculations, except recovery of the tracer is quantitative and no losses occur when cutting the core plugs. Tests for formation damage conducted with Point McIntyre cores indicate tracer recovery from the core is quantitative. The Dean-Stark water extraction recovered 99.6% of the tracer. Contamination of cores by core-plugging fluid was 18.1 and 18.7% of the oil-phase weight in two tests. These laboratory tests indicate there is some replacement of in-place oil by core-plugging fluid, but 75% of the oil originally present remained in the core after plugging.

The contract core-analysis laboratory should provide a spreadsheet listing all weights measured during the core analysis. *Note: The tracer could be found in the side arm and also in the toluene in the core. The retained toluene volume in the core is usually a small fraction of the total toluene used in the Dean-Stark water extraction; therefore, it is not a significant error.*

Appendix D

Calcium Carbonate Determination in Whole Mud

There may be times when it is desired to determine the amount of calcium carbonate in WBM. One example from a past project involves the scenario of two core points 1,000 ft apart. The coring fluid composition may be optimal for the first core, but after drilling the next 1,000 ft, half of the original calcium carbonate may be depleted. This appendix contains a method for determining the amount of calcium carbonate in whole WBM. It is simple, accurate, and uses only chemicals already available at the rig. The procedure involves using hydrochloric acid to dissolve the calcium carbonate, and then titrating for the soluble calcium. Because OBMs may contain high concentrations of calcium chloride, this method has not yet been adapted to determine the amount of calcium carbonate in an OBM. It may be possible to break the OBM emulsion as in the normal API test, and filter out the solids (including the calcium carbonate) before running the test described in this appendix.

1. Pour approximately 40 cm^3 of deionized (or distilled) water into a 400-mL beaker, add a magnetic stirring bar, and then place the beaker on a stir plate.
2. Precisely measure 10 cm^3 of whole mud into the beaker while stirring. *Note: A large syringe may be the easiest measuring device. (If measuring with either a graduated cylinder or similar measuring device, rinse all the mud into the beaker with 50 cm^3 of either deionized or distilled water.)*
3. While stirring, carefully add 25 cm^3 of 15% hydrochloric acid (HCl) to the beaker. The HCl dissolves the calcium carbonate. *Note: Be sure to add the acid very slowly at first to control the resulting effervescence from becoming uncontrollable.* Stir the solution for at least 5 minutes.
4. Pour the solution into a 200-mL volumetric flask. Rinse the 400-mL beaker with either deionized or distilled water, then add that liquid to the flask. Continue rinsing until all the material has been rinsed into the flask, and then add enough water to bring the total liquid in the flask exactly to the 200-mL mark. Cap the flask, and shake the solution to be sure it is homogeneous.
5. Transfer exactly 2 cm^3 of solution from the flask into a titrating dish. Add a stirring bar. Add 2 cm^3 of 1.0 N NaOH and 0.5 g of Calver II indicator.
6. While stirring, titrate from red wine color to blue with Versenate hardness titrating solution (1 mL = 20 equivalent parts per million Ca).
7. Calculate the ppb of calcium carbonate in the mud as follows:

   (cm^3 of titrating solution) × (3.5) = ppb calcium carbonate.
   Example: 15.7 cm^3 × 3.5 = 55 ppb calcium carbonate.

Appendix E

Iodide Ion Analysis

A primary goal of low invasion coring (LIC) is to determine the original water saturations in cores so that hydrocarbon reserves can be inferred. Water from coring fluids can contribute to the total water found in core samples, and the use of tracers such as iodide salts can make it possible to correct for this contamination. Complete extraction of the iodide salt can be accomplished only by crushing the core samples, so this step is the last step and is performed after all other core analyses are complete. The iodide salt extraction and analysis will be performed by the contract core-analysis laboratory, or by another analytical laboratory designated by the operator.

1. Use the routine horizontal core plug (not extracted with methanol) for chloride ion analysis.
2. Redry the core plug. Allow the core plug to cool, record the dry weight, and then compare the weight to the original routine dry weight.
3. Crush the core plug using a mortar and pestle. Transfer the crushed material to a tared flask, and then weigh. Calculate the amount of crushed rock in the flask.
4. Rinse the mortar and pestle with distilled water, and then pour the water into the flask. Add 1.5 cm^3 of distilled water per gram of rock into the flask. Weigh and record the total weight of distilled water added to the flask.
5. Extract the sample of ground core material with continuous shaking overnight (8 hours) using a wrist shaker.
6. Filter some of the supernatant liquid (at least 20 cm^3) through a 0.45-μm cellulose nitrate cartridge filter. Analyze the liquid for iodide using ion chromatography.
7. Analyze blank samples of distilled water for iodide using ion chromatography.
8. Analyze the iodide tracer concentration in the mud samples using ion chromatography. Record the mud-sampling depths. Correlate the tracer concentration to the sample depths.

   Note:

   a. Iodide concentration in the formation brine is usually below 1 ppm and can be ignored. Whenever possible, the actual iodide concentration in the formation brine should be verified. If the iodide ion concentration in the formation brine exceeds 1% of the iodide ion tracer concentration, then subtract the formation brine concentrations from the final iodide ion totals.

   b. The same sample can be used for the chloride analysis (see Appendix H).

9. Calculate the concentration of iodide in the Dean-Stark water volume using the following equation:

$$C_I^{ds} = C_I^{\text{flask}} \frac{W_w^{\text{flask}}}{W_w^{ds}\left(\dfrac{W_r^{\text{flask}}}{W_r^{ds}}\right)}.$$

10. Calculate the weight fraction of mud filtrate in the Dean-Stark water volume as follows:

$$F_{mf} = \frac{C_I^{ds} - C_I^{cw}}{C_I^{mf}}.$$

11. Calculate the connate water saturation using the following equations:

$$V_{cw}^{+} = \frac{W_w^{ds}}{\left(\frac{10^6 - C_s^{cw}}{10^6}\right)\rho_{cw}}$$

$$V_{cw} = V_{cw}^{+}(1 - F_{mf})$$

$$S_{cw} = \frac{V_{cw}}{PV_r}.$$

12. Report the mud-filtrate saturation and calculated connate water saturation in the final routine core analysis.

Definition of Terms

C_I^{cw} = iodide concentration in connate water, ppm
C_I^{ds} = iodide concentration in Dean-Stark water volume, ppm
C_I^{flask} = iodide concentration in distilled water, ppm
C_I^{mf} = iodide concentration in mud filtrate (correlated to plug sample depths), ppm
C_s^{cw} = concentration of salts, ppm
F_{mf} = weight fraction of mud filtrate in Dean-Stark water volume
PV_r = PV of the rock, cm^3
S_{cw} = connate water saturation, fraction PV
V_{cw} = volume of connate water, cm^3
V_{cw}^{+} = volume of connate water with contaminants (mud filtrate), cm^3
W_r^{ds} = weight of dried core plug after Dean-Stark water extraction, g
W_w^{ds} = weight of Dean-Stark water volume, g
W_r^{flask} = weight of crushed rock transferred into tared flask, g
W_w^{flask} = weight of distilled water added into the tared flask, g
ρ_{cw} = density of connate (formation) brine, g/cm^3

Appendix F

Bromonaphthalene-Analysis Procedure

F.1 Introduction

Selected routine core plugs and off-site routine core plugs are usually analyzed for tracer to quantify mud-filtrate invasion, if any. Selected routine core-plug endpieces (i.e., zones with highest permeability and low ROP) can be analyzed for a tracer to estimate the depth of mud-filtrate invasion. The drilling mud filtrates should be analyzed to determine a baseline concentration to be used in the calculations related to mud-filtrate invasion. Each core sample is extracted with approximately 125 mL of reagent-grade toluene. The degree of mud-filtrate invasion by the coring-fluid filtrate is based on the concentration of tracer measured in the toluene extract and the PV of the extracted core plug. In a similar fashion, the cutting-fluid oil invasion is based on the concentration of the cutting-fluid oil in the toluene extract.

The 1-bromonaphthalene is usually the tracer for OBM. If the base oil has a unique gas chromatography (GC) fingerprint, it may be used as a backup tracer; but its maximum sensitivity is only approximately 10% PV of mud-filtrate invasion, which can be used in the event of a problem with the analysis for 1-bromonaphthalene. GC with an electron-capture detector (ECD) is the proper choice to quantify the 1-bromonaphthalene. A flame-ionization detector (FID) is a good choice for quantifying the base-oil concentration, if needed. A splitter before the detector is used in the analysis for the ECD and FID analyses to be generated simultaneously. An injection-port splitter is used to separate the toluene, so rotary evaporation (to concentrate the sample) is not required to analyze for base oil or plugging oil. The base-oil analysis is only qualitative, but can be made quantitative, if needed. A qualitative indication of base-oil concentration is usually printed on the data sheet with a scale of 0 (i.e., no base oil detected) through 3 (i.e., high base oil detection). The procedures for the quantitative analyses of 1-bromonaphthalene and the plugging oil are presented in this appendix. A contract tracer-analysis laboratory usually performs these chromatographic analyses, but they can be done in-house if the capability exists. Tightly capped and taped containers of the toluene are transferred to that location by the contract core-analysis laboratory.

F.2 Scope

This procedure provides accurate quantification of drilling mud and core-plug cutting-fluid invasion in core plugs. To accomplish this quantification, a halogenated tracer (1-bromonaphthalene) is added to the drilling fluid at a concentration of 100 ppm. Each core sample is extracted with approximately 125 mL of reagent-grade toluene. The subsequent core extracts are analyzed by GC using an ECD and an FID. The degree of core invasion by the drilling-mud filtrate is based on the concentration of tracer contained in the toluene extract and the PV of the extracted core plug. In a similar fashion, the cutting fluid (Isopar) invasion is based on the concentration of the Isopar in the toluene extract.

Equipment models may change with time, but this procedure is a good starting point for the development of an acceptable analytical procedure.

F.3 Instrumentation

- HP-5880A gas chromatograph—Level 4
- HP Model 7364 autosampler
- HP Model 19312 nickel 63 ECD
- PE/Nelson 25 m × 0.53-mm ID—0.5-m film #N931-2689
- Valco three-port valve, high temperature
- Valco 1/16-in. 304 stainless steel tee
- HP autosampler vials and caps
- 1-mL glass disposable pipettes or equivalent

F.4 Reagents

- 1-Bromonaphthalene; Eastman Part #2256 or equivalent
- Reagent-grade toluene
- 1-Iododecane; Aldrich Part #23,825-2 or equivalent
- 2-ppm iododecane and 1.0 % squalane in toluene
- Standards of 1-bromonaphthalene (0.1 to 10.0 ppm) in toluene and crude oil (20:1)
- Field stock-tank crude oil
- Squalane; Aldrich Part #23,431-1 or equivalent

F.5 GC Conditions

Oven Profile

Initial temperature	= 75°C
Initial time	= 2.00 minutes
Program rate	= 20°/min
Final temperature	= 300°C

Detector and Injector

Detector temperature	= 325°C
Injector temperature	= 300°C
Detector attenuation	= 2^8
Injection volume	= 5.0 μL

Gas Flows

Column flow	= 9.5 cm^3/min
Nitrogen makeup	= 100 cm^3/min
Splitter flow	= 50 cm^3/min
Septum purge	= 3.5 cm^3/min

F.6 Analytical Procedure

1. On receipt from the core-analysis laboratory, each sample bottle of tracer and toluene is weighed to ±0.1 g, and each weight is recorded in a log book. *Note: The samples are refrigerated at all times when not being processed.*
2. First, a 0.5-mL aliquot of each sample is diluted 1:1 by volume with the internal standard solution (i.e., 2-ppm iododecane and 1.0% squalane in toluene).

Note: In the case of the concentrated drilling-fluid solutions, it is necessary to predilute them ⊕0.4 g of sample to 10.00 g with reagent-grade toluene. Actual weights are recorded to ±0.001 g, along with the respective dilution factor.

3. The 5.0 μL of the sample/internal standard blend is injected into the GC splitter by means of the autosampler, and then the data acquisition is initialized.
4. After the internal standard and tracer have eluted, the three-port valve is actuated, and the oven temperature is allowed to proceed to its maximum setting.
5. From the ECD, the concentration of the tracer is determined using the 1-ppm iododecane peak as an internal standard, and also using the response factor relating the area counts to concentration. The internal standard is used to correct for variations in sample size reaching the detector.
6. From the FID, the concentration of the core-plug cutting fluid is determined using the 0.5% squalane peak as an internal standard, and also using the response factor relating the area counts to concentration.
7. The measured tracer concentration is to be adjusted for any dilutions and then reported as ppm tracer.
8. The measured core-plug cutting fluid concentration is to be adjusted for any dilutions and then reported as percent cutting fluid.
9. A three-point calibration curve of tracer in toluene and crude oil is run daily. Response factors are calculated from the calibration curve, and a running log of the calculated response factors vs. time is maintained daily.

F.7 Tracer Calculations

1. Drill the sample from the whole core using either brine or humidified gas as the bit coolant.
2. Weigh the clean and dried Dean-Stark flask. Complete the Dean-Stark water extraction of fresh-state core plug with toluene.
3. Weigh and calculate the total amount of toluene with contaminants (e.g., crude oil) after the toluene extraction (add side arm toluene to the flask before measuring the weight).
4. Analyze the tracer concentration in the mud samples. Record the mud-sampling depths. Correlate the tracer concentration to the sample depths.
5. Calculate the total weight of the oil plus contaminants (e.g., mud filtrate) in the core before the Dean-Stark water extraction as follows:

$$W_o^+ = W_r - W_r^{ds} - W_w^{ds}.$$

6. Calculate the mass of the mud filtrate in the core as follows:

$$W_{\text{tracer}}^r = C_{\text{tracer}}^{ds} \times W_t^{ds}$$

$$W_{mf}^r = \frac{W_{\text{tracer}}^r}{C_{\text{tracer}}^{mf}}.$$

7. Calculate the saturation of the oil in the core using the following equation:

$$S_o^r = \frac{W_o^+ - W_{mf}^r}{\rho_o \times PV_r}.$$

The contract core-analysis laboratory should provide a spreadsheet listing all weights measured during the core analysis. *Note: The tracer can be found in the side arm and also in the toluene in the core. The retained toluene volume in the core is usually a small fraction of the total toluene used in the Dean-Stark extraction; therefore, this is not a significant error.*

All calculations using concentration need to be converted from ppm to weight fraction as follows:

$$\text{wt fraction} = \text{ppm}/10^6.$$

F.8 Definition of Terms

C_{tracer}^{mf} = tracer concentration in the mud filtrate, ppm

C_{tracer}^{ds} = tracer concentration in the Dean-Stark toluene, ppm

PV_r = PV of the core, cm^3

S_o^r = saturation of the oil in the core, PV fraction

W_{mf}^r = weight of the mud filtrate in the core, g

W_o^+ = weight of oil plus contaminants (e.g., mud filtrate) in the core before Dean-Stark water extraction, g

W_r = weight of fresh state core before Dean-Stark water extraction, g

W_r^{ds} = weight of dry core after Dean-Stark water extraction, g

W_t^{ds} = weight of toluene and plus contaminants (e.g., mud filtrate) after Dean-Stark water extraction, g

W_{tracer}^r = weight of tracer in the core, g

W_w^{ds} = weight of Dean-Stark water volume, g

ρ_o = density of the oil, g/cm^3

Appendix G

Sample Anaerobic-Crude-Oil Collection Letter

Note: This letter would be sent by the Coring Project Manager (or designee) to field personnel where the coring will take place.

Date: File No. ###
Subject: **Anaerobic Crude-Oil Sample**
From/Location:
Telephone:
To/Location:

The containers for the anaerobic crude-oil samples should be evacuated before use, and labeled as empty for shipping.

The oil/water/gas separator from which the samples are collected should be thoroughly flushed with oil from the well to be sampled. All chemical treatments at the separator, such as emulsion breakers and corrosion inhibitors, should be terminated before flushing of the separator begins. Normally, all well treatments should be terminated by skipping one or two treatment cycles before conducting the test. When a new well is being tested, the fluids should be pristine. Therefore, it is expected that well treatments will not be an issue in this situation.

Estimate the oil residence time of the separator. Consider the separator flushed when the oil volume is displaced five times. An alternative approach is to drain the separator, and then refill it with the fluids to be sampled.

The cylinders are rated to 240 psia. *Note: Never exceed this pressure.* If the separator operates above 200 psia, connect a pressure gauge and valve on top of the cylinder, and install a throttling valve between the bottom cylinder valve and the separator sampling port. Use the valve to control flow into the cylinder maintaining the pressure below 200 psia. *Note: Do not use the cylinder on/off valve or the separator sample port on/off valve as the throttling valve.* You may want to include the pressure gauge even if the separator operating pressure is less than 200 psia.

Frequently the cylinders are filled at the completion of the well test with the separator shut in and the well diverted to the normal production manifold. Separator pressure then can be reduced to minimum operating levels and the cylinders filled. Using this approach minimizes the volume of gas to be vented and should reduce sampling time.

Connect the sample container (cylinder) to a sampling port on the separator. The cylinder connection should be ¼-in. NPT (National Pipe Thread). Purge all air from the line between the separator and the valve by bleeding the oil from the line before the connection is tightened. Fill the cylinder from the bottom, and then bleed the gas from the top. The gas should be vented into a well-ventilated area. If the separator is in an enclosure, a line should be connected to the outlet valve to carry the vented gas outside the enclosed area.

Crude oil tends to foam as the gas breaks out of solution. Therefore, oil may appear at the outlet valve prematurely, if the filling process is conducted too rapidly. A 20% gas space must be left in the cylinder to allow for potential thermal expansion during shipment. You can estimate the required voidage by filling the cylinder only 80% full. A more precise approach (i.e., the preferred approach) is to fill the cylinder and then drain 2 gal from the cylinder after filling is completed and the cylinder is disconnected from the separator. Next, bleed the pressure from the cylinder, creating a stock-tank crude-oil sample. A residual pressure of a few psig is acceptable. Screw the plugs into the valves. Label each cylinder with the date, well number, filling temperature, final cylinder pressure, and then void the volume. Ship the cylinders to the following:

[Laboratory Name]

[Operator Name]

[Laboratory Address]

If you have any questions, please contact me at [PHONE NUMBER/EMAIL ADDRESS].

Appendix H

Chloride Ion Analysis—Brine Saturation

Determining the salinity of the brine in a core is critical to the accurate calibration of logging tools, and can help identify waterflood breakthroughs. Complete extraction of the salts can be accomplished only by crushing the core samples, so this step is performed last after all other core-analysis measurements are completed. The salt extraction and chloride analysis will be performed by the contract core-analysis laboratory, or by an analytical laboratory designated by the operator.

H.1 Procedure

1. Use the routine horizontal core plug (not extracted with methanol – salts are somewhat soluble in methanol) for chloride ion analysis.
2. Redry the core plug. Allow the core plug to cool and record the dry weight—compare with the original routine dry weight to ensure the core is completely dried.
3. Crush the core plug using a mortar and pestle. Transfer crushed material to a tared flask, and then weigh. Record the weight of crushed rock in the flask.
4. Rinse the mortar and pestle with distilled water, and then pour the water into the flask. Add 1.5 cm^3 distilled water per gram of rock into the flask. Weigh and calculate the total amount of distilled water added to the flask.
5. Extract the water soluble salts with continuous shaking overnight using a wrist shaker.
6. Filter the supernatant liquid (at least 20 cm^3) through a 0.45-μm cellulose nitrate cartridge filter. Analyze the filtered liquid for chloride using ion chromatography.
7. Analyze blank samples of distilled water for chloride using ion chromatography.
8. Analyze the chloride concentration in the mud samples using ion chromatography. Record the mud-sampling depths. Correlate the chloride concentration to the sample depths.
9. Calculate the concentration of chloride in the Dean-Stark extracted water volume as follows:

$$C_{Cl}^{ds} = C_{Cl}^{\text{flask}} \frac{W_w^{\text{flask}}}{W_w^{ds}\left(\dfrac{W_r^{\text{flask}}}{W_r^{ds}}\right)}.$$

10. Calculate the weight fraction of mud filtrate in the Dean-Stark core-plug water volume as follows:

$$F_{mf} = \frac{C_{Cl}^{ds} - C_{Cl}^{cw}}{C_{Cl}^{mf}}.$$

11. Calculate the connate water saturation using the following equations:

$$V_{cw}^{+} = \frac{W_{w}^{ds}}{\left(\frac{10^6 - C_{s}^{cw}}{10^6}\right)\rho_{cw}}$$

$$V_{cw} = V_{cw}^{+}(1 - F_{mf})$$

$$S_{cw} = \frac{V_{cw}}{PV_{r}}.$$

12. Report the mud-filtrate saturation and calculated connate water saturation in the final routine core analysis data spreadsheets/tabulations.

H.2 Definition of Terms

C_{Cl}^{cw} = chloride concentration in connate water volume, ppm

C_{Cl}^{ds} = chloride concentration in Dean-Stark water volume, ppm

C_{Cl}^{flask} = chloride concentration in distilled water, ppm

C_{Cl}^{mf} = chloride concentration in mud filtrate (correlated to plug sample depths), ppm

C_{s}^{cw} = concentration of salts, ppm

F_{mf} = weight fraction of mud filtrate in Dean-Stark water volume

PV_{r} = PV of the rock, cm^3

ρ_{cw} = density of connate (formation) brine, g/cm^3

S_{cw} = connate water saturation, fraction PV

W_{r}^{ds} = weight of dried core plug after Dean-Stark water extraction, g

W_{w}^{ds} = weight of Dean-Stark water volume, g

W_{r}^{flask} = weight of crushed rock transferred into the tared flask, g

W_{w}^{flask} = weight of distilled water added into the tared flask, g

V_{cw} = volume of connate water, cm^3

V_{cw}^{+} = volume of connate water with contaminants (e.g., mud filtrate), cm^3

Appendix I

Impact of Oil-Based Mud on Connate Water Saturation and Other Rock Properties

I.1 Background

During coring, oil-based mud (OBM) filtrate invades the core, either partially or totally. The OBM filtrate contains surfactants required to give the mud some of its desired suite of rheological and solid-carrying properties. Some of the mud components are weak surfactants, and others are considered strong surfactants. There is concern that these surfactants can alter either part or all of the following: rock wettability, water/oil relative permeability relationships, rock electrical properties, and connate water saturation (S_{wc}).

This appendix discusses the potential for mobilizing S_{wc} based on studies previously published in the technical literature and discussed in this appendix (Abrams 1975; Dombrowski and Brownell 1954). The discussion presented herein concerns not only the interfacial tension (IFT) between OBM filtrate and connate brine, but also the pressure gradients within the core as it is being cut, and the limited time the core is exposed to these pressure gradients. While the wettability of portions of the pore system may be altered by OBM-filtrate invasion, the mobilization of connate water requires not only a very low-IFT environment, but also a sufficient flow rate (or pressure gradient). Using terminology from enhanced-oil-recovery (EOR) technology, the degree of mobilization of an immobile phase is dependent on a combination of the flow rate and the IFT as typically combined into a dimensionless capillary number (N_c).

This appendix is a brief review and discussion of key concepts regarding properties of OBM filtrate, the time-dependent relationship of the IFT of the OBM-filtrate/connate-water fluid pair, and the pressure gradients within freshly cut rock at the coring bit. The two most relevant SPE technical papers are Reed et al. (1994) and Warner and Rathmell (1997), and they are listed and summarized in Appendix A. For a more in-depth review of this topic, the reader is referred to the list of references at the ends of these two technical papers.

Note: The purpose of the procedures presented in this monograph is to reduce or eliminate OBM-filtrate invasion during coring. This appendix discusses the consequence if such OBM-filtrate invasion occurs.

I.2 Relationships of Mud-Filtrate/Connate Brine IFT and Original Reservoir Properties

The effects of OBM-filtrate invasion on the reservoir cores' S_{wc} and wettability can be difficult to define and are not discussed extensively in the technical literature. During the past several decades and for obvious economic reasons, the oil industry, for the purposes of EOR and mobilizing residual oil saturation, extensively studied the relationship of residual oil saturation to water/aqueous phase displacement (S_{orw}) vs. IFT, flow velocity, and other relevant parameters. In particular, **Fig. I.1** shows the semilogarithmic relationship determined experimentally between S_{orw} and N_c for several rock

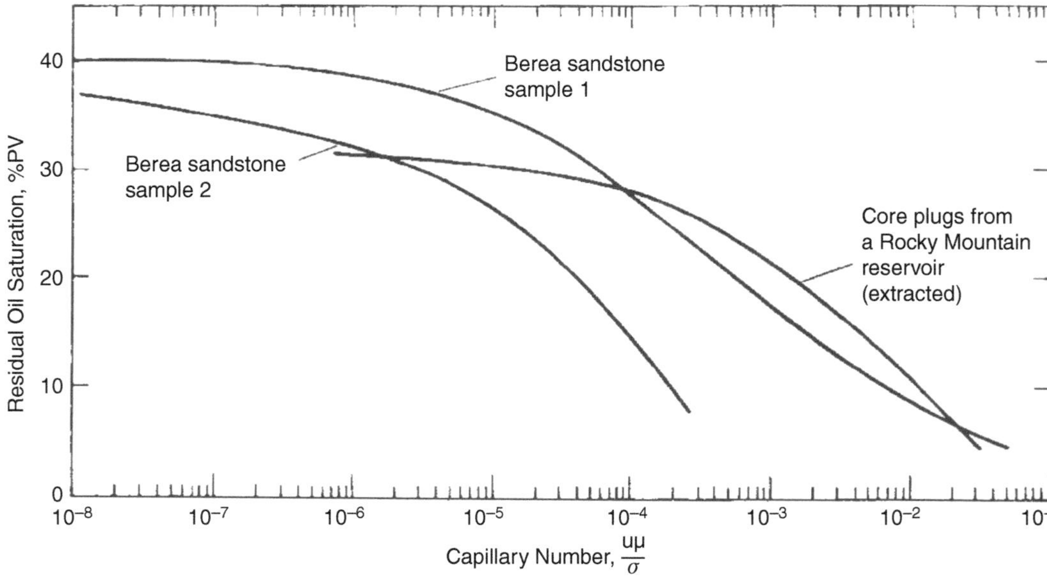

Fig. I.1—Dependence of residual oil saturation (S_{or}) on capillary number (N_c). From Stalkup (1983).

types based on water-displacing-oil laboratory experiments (Stalkup 1983). N_c is equal to $u\mu/\sigma$ where u is flow velocity, μ is viscosity, and σ is the IFT. *Note: On Fig. I.1, the curved relationship requires order-of-magnitude changes in the N_c value to change the S_{orw} significantly.* Because the IFT is in the denominator of the N_c, large IFT values lead to small N_c values and vice versa; therefore, lower IFT values (particularly changes in orders of magnitude) are needed to mobilize much S_{orw}. Abrams (1975) also published similar data, as seen in **Fig. I.2.**

The situation for S_{wc} is different from that of S_{orw} because the portions of the pore space occupied by the oil and water immobile phases are so different. In the S_{orw} situation, the disconnected oil globules are primarily in the larger pore spaces (pore spaces having much of the reservoir's permeability). In the situation concerning S_{wc}, the water is located in the smallest pore spaces, nooks, and crannies. This is because the rock, when deposited, was originally filled with water and was water-wet; and then later, as the oil (or gas) filled the reservoir, the hydrocarbon-filling process displaced the water from the pore spaces. This displacement occurs first in the largest pores and continues into smaller and smaller pore spaces, which means that at field discovery the connate water is located in the smallest portion of the reservoir pore-size distribution. The situation regarding wettability is more complicated because the wettability of the large pore spaces can be different from that of the fine pore spaces; and even in the large pore spaces, some of the rock surfaces may be oil-wet and other portions water-wet (mixed wettability).

Dombrowski and Brownell (1954) published a significant study concerning the mobilization of the residual aqueous wetting-phase saturation as a function of N_c. This experimental study used a variety of unconsolidated porous media and a wide range of IFT values and flow rates (or pressure gradients). **Fig. I.3** is the key figure from their paper. The curve in Fig. I.3 has a shape similar to those of Fig. I.1 and Fig. I.2; however, it is shifted to the right by approximately two orders of magnitude. This shift means that it takes a much lower IFT value, or a much higher flow rate, to mobilize the residual wetting phase saturation, generally the connate brine, than it does to mobilize the residual nonwetting phase saturation, generally oil.

The technical literature contains only a few studies about the relationship of S_{wc}, OBM-filtrate/connate-brine IFT, and flow velocities (or pressure gradients). Two of these papers are discussed herein [Reed et al. (1994) and Warner and Rathmell (1997)]; both papers concern coring with OBM.

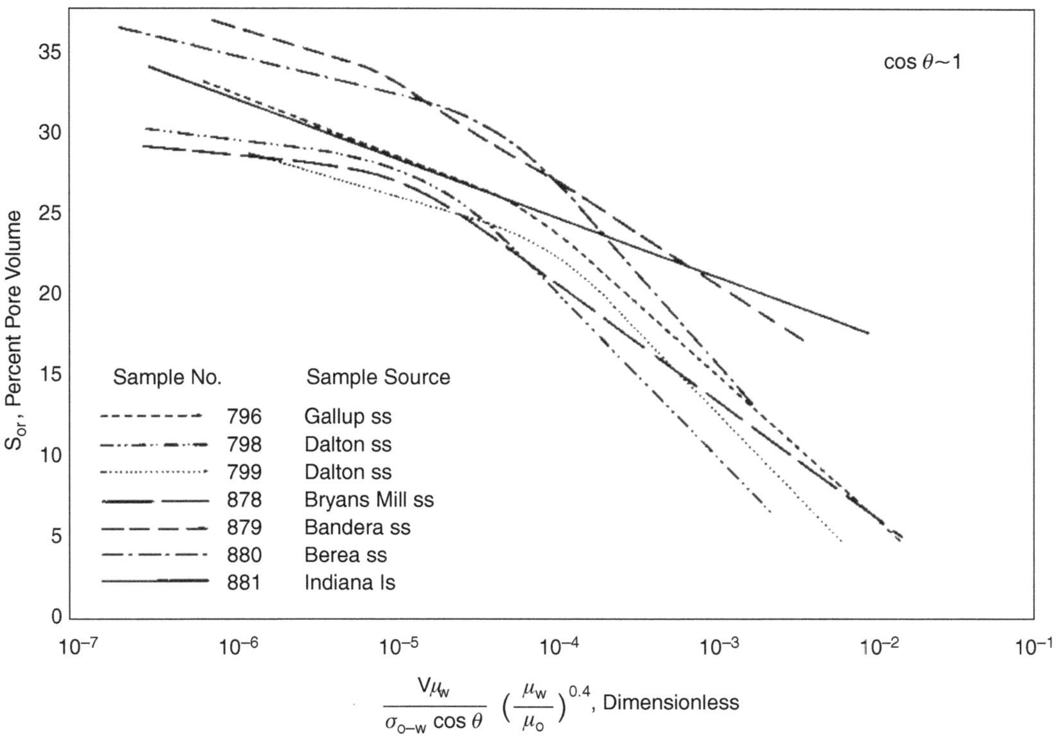

Fig. I.2—Effect of fluid viscosity, IFT, and flow velocity on S_{or} for various rock samples (from Abrams 1975).

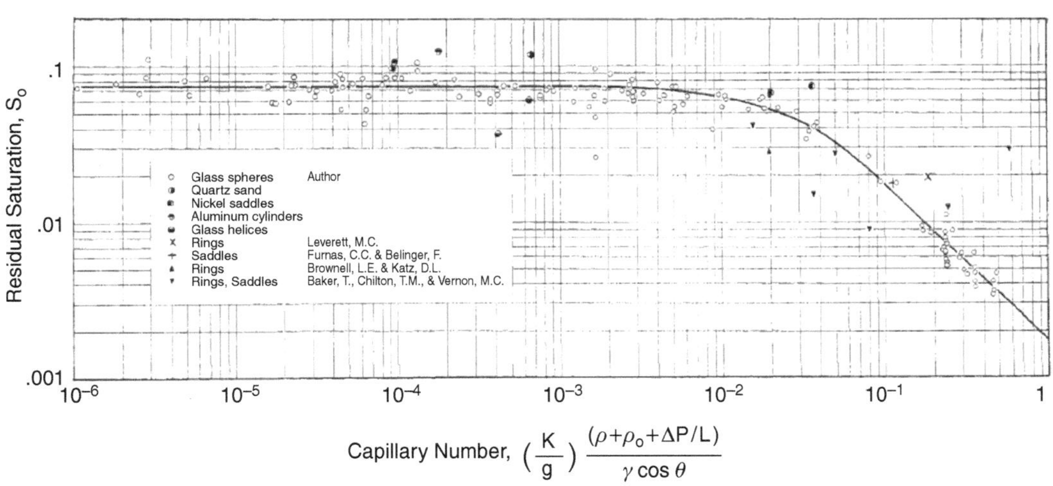

Fig. I.3—Residual saturation vs. capillary number for all unconsolidated porous media. Reprinted with permission from Dombrowski, H.S. and Brownell, L.E. Residual Equilibrium Saturation of Porous Media. ***Industrial and Engineering Chemistry*** **(June 1954): 1207–1219. Copyright 1954 American Chemical Society.**

These papers are based on the use of a high-quality OBM with very low water content. The papers show that the methodology and procedures for experimental studies of the interaction between OBM filtrate and S_{wc} must be developed very carefully because it is quite easy to impose conditions on the laboratory experimental setup that not only do not replicate reservoir conditions, but that also can produce results opposite of those in actual reservoir environments.

Reed et al. (1994) and Warner and Rathmell (1997) do not address the effect of OBM filtrate on reservoir rock wettability, relative permeability, or electrical properties. These are more difficult topics because (a) establishing the baseline from which to measure any changes is very difficult and (b) the initial and boundary conditions of the laboratory experimental systems easily impact the outcome of the experiments.

Reed et al. (1994) describe laboratory experiments—namely IFT measurements and core-flushing studies—performed to determine the effect that certain experimental variables have on the effect of OBM-filtrate invasion on resident water saturations. "Laboratory evaluations of the potential for OBM filtrate to displace connate water during the coring process included (1) the measurement of interfacial tensions (IFT) between whole mud, various mud filtrates and both gas-cap and aquifer brines, and (2) a series of core-flushing tests using fluids similar to those in the IFT work. Transient and equilibrium IFT measurements were conducted using sessile-drop equipment. The tests were performed at varying temperatures and under a 500-psi CO_2 pressure to approximate reservoir conditions." *Note: In this experimental study, CO_2 was included in the hydrocarbon mixture because the Prudhoe Bay Ivishak reservoir hydrocarbon gas contains approximately 10% CO_2.*

Fig. I.4 (Reed et al. 1994, Fig. 3) shows the IFT values between the OBM-filtrate/brine-fluid pair as a function of time (hours), and the short-time equilibrium IFT values are quite different from, and higher than, those measured for long periods of time.

In this series of laboratory flow experiments, carboxylate concentrations ranged from 0.34 to 1.89 wt% based on the content in whole mud and mud centrifugate. Pressure gradients applied during the flow test varied from 5 to 250 psi/in. However, the carboxylate concentrations measured from actual field cores by core-layering experiments ranged from approximately 0.9 to 0.02 wt%, as shown in **Figs. I.5 and I.6** (Reed et al. 1994, Figs. 1 and 2).

The results of these flow tests are presented in **Table I.1** (Reed et al. 1994, Table 6) and show very little water movement under almost all but the most extreme experimental conditions.

Regarding capillary numbers, the authors state that: "Table 5 (**Table I.2** in this appendix) shows the capillary numbers that result for various combinations of interfacial tensions, permeabilities, and pressure gradients. Based on the previous discussion of IFT measurements, the most likely IFT in the core is 1 dyne/cm or greater. The pressure gradient in the core, based on computer simulation of the coring, is in the range of 1 to 5 psi/in. These values, combined with 1,000-md permeability, yield an N_c of $1.4{\times}10^{-3}$. The Dombrowski-Brownell (1954) data indicate that an N_c higher than 10^{-2} is

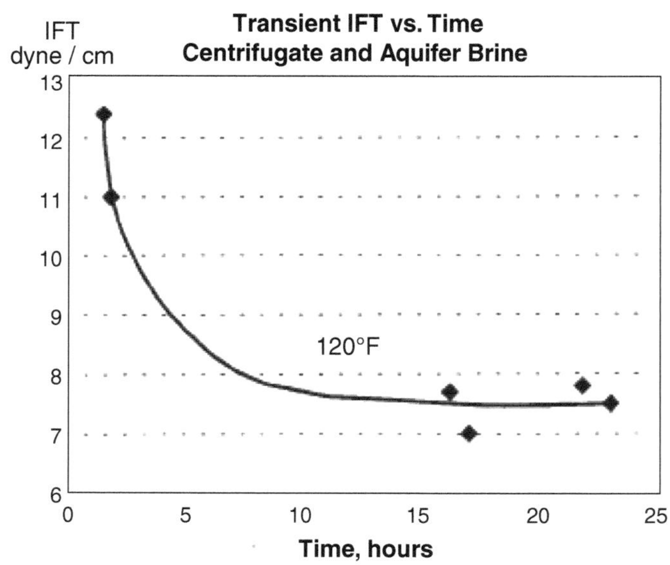

Fig. I.4—Transient IFT (dynes/cm) vs. time (hours) for OBM centrifugate/aquifer brine fluid pair. From Reed et al. (1994).

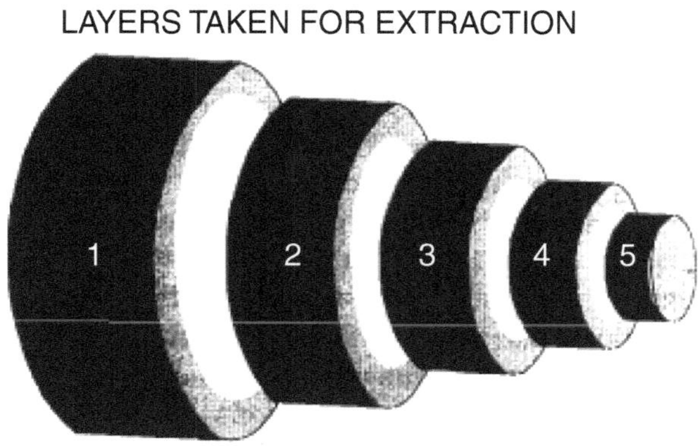

Fig. I.5—Schematic of five core layers subjected to carboxylates determination in core-layering experiments. From Reed et al. (1994).

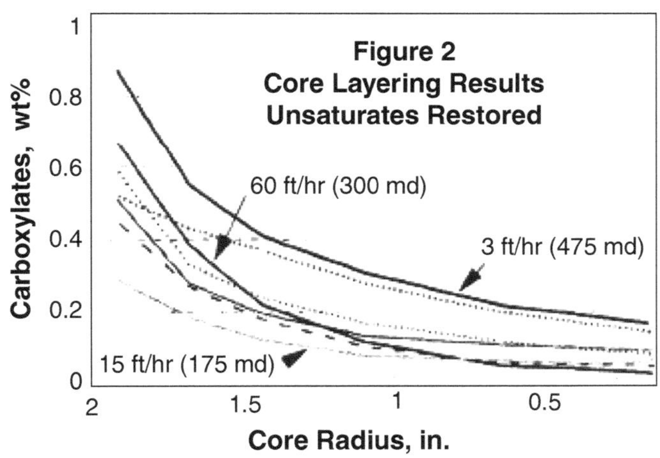

Fig. I.6—Carboxylate concentration as a function of distance from core center (core center @ radius = 0 in., and 4-in. core outer edge @ radius = 2 in.) for core samples cut at three different coring rates (3, 15, and 60 ft/hr). From Reed et al. (1994).

TABLE I.1—S_w FOR VARIOUS FLUSHING TESTS (FROM REED ET AL. 1994)

Test	*P* (psi)	CARBOX wt%	CO_2	S_{wi} (%PV)	S_w	S_{wi}/PV_i
A1/2	500	1.89	N	28.2	12.8	2.50
B1/2	80	1.89	N	29.2	7.2	0.60
B3/4	500	1.89	Y	25.8	1.9	0.10
B5/6	500	0.49	N	25.4	1.2	0.02
C1	100	0.34	Y	16.8	0.1	0.05
C2	95	0.98	Y	14.2	0.6	0.05
C3	201	0.82	Y	14.2	0.5	0.02
C4	11	0.91	Y	17.3	0.3	0.01
C5	12	0.91	N	18.1	0.2	0.03

TABLE I.2—CAPILLARY NUMBERS—POSSIBLE RANGE OF CORING CONDITIONS (FROM REED ET AL. 1994)

	$N_c \times 10^3$			
	IFT = 1 dyne/cm		IFT = 0.1 dyne/cm	
Pressure Gradient	400 md	1,000 md	400 md	1,000 md
5 psi/in.	0.54	1.4	5.7	14
10 psi/in.	1.1	2.7	11	28

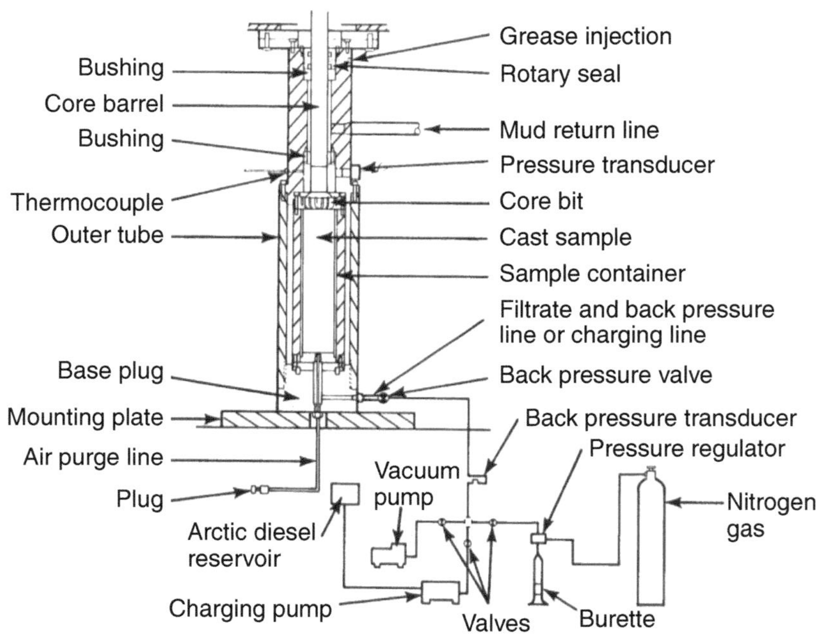

Fig. I.7—Schematic of Drilling Research Laboratory coring equipment setup used for minicoring tests. From Warner and Rathmell (1997).

needed for displacement of water. The Omoregie data show no displacement with N_c as high as 10^{-1} if the capillary number is assumed to equal one-half the Bond number. If bounding limits of 0.1 dyne/cm, 10 psi/in., and a 1,000-md permeability are assumed, $N_c = 2.8 \times 10^{-2}$ and water displacement becomes possible based on Dombrowski-Brownell" (Reed et al. 1994).

The conclusions of Reed et.al. (1994) include the following:

- "All transient measurements and all equilibrium tests with aquifer brine resulted in IFTs that indicate no likelihood of connate water displacement by OBM filtrate invasion.
- "The state of equilibrium between resident brine and invading filtrate remains speculative. Coring rates varied from 4 to 60 ft/hr, and the length of core exposed to dynamic filtration (cutting edge of the bit) was 4 in. Time of dynamic filtration can then be calculated to be 5 minutes or less. The transient IFT tests show that several hours may be required to reach equilibrium values.
- "Tests show water saturation reductions are likely when carboxylate contents approach 2 wt%, there is a high-pressure gradient, and CO_2 is absent.
- "Tests B-3 through B-6 demonstrated little or no water is removed from the core at high-pressure gradients and high-carboxylate contents if CO_2 is present, and at high-pressure gradients if the carboxylate content is reduced to the 0.5 wt% level.

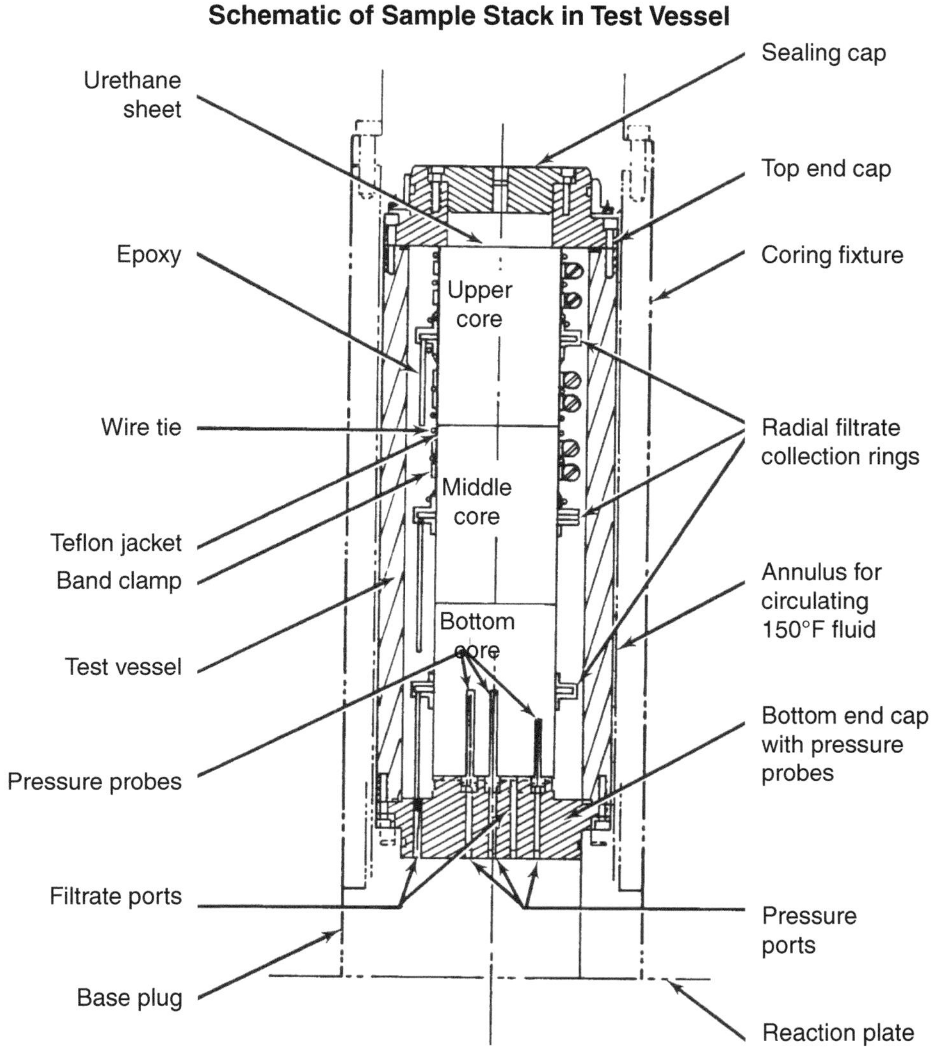

Fig. I.8—Schematic of DRL coring equipment setup used for minicoring tests. From Warner and Rathmell (1997).

- "All C Series tests show insignificant water removal. These results imply that either the low IFTs measured by Reed in his gas-cap brine equilibrium test are not established in the core or a capillary number greater than 2.8×10^{-2} is needed to result in water displacement."

Warner and Rathmell (1997) present the results of a series of minicoring experiments in which 4-in. cores are cored into 2-in. minicores [see **Fig. I.7** (Warner and Rathmell 1997, Fig. 2) for a schematic of the experimental setup]. The aspect of these experiments most revealing is the pressure data obtained from three embedded pressure taps: one at the center of the core, one located just within the 2-in. diameter of the mini-core, and one located toward the edge of the 4-in. core so that the coring bit directly drilled into this pressure tap [see **Fig. I.8** (Warner and Rathmell 1997, Fig. 1)].

Some of the results of these experiments are presented in **Figs. I.9 and I.10.** Particularly, the pressure vs. penetration plots show almost all the pressure differential between the mud system and the pore pressure is taken up in the mudcake, even in the dynamic zone directly beneath the coring bit.

The authors state: "As the coring bit approaches pressure tap P1, the pressure gradient in the rock ahead of the coring bit to the bottom of core stack and at the core center can be determined. Based

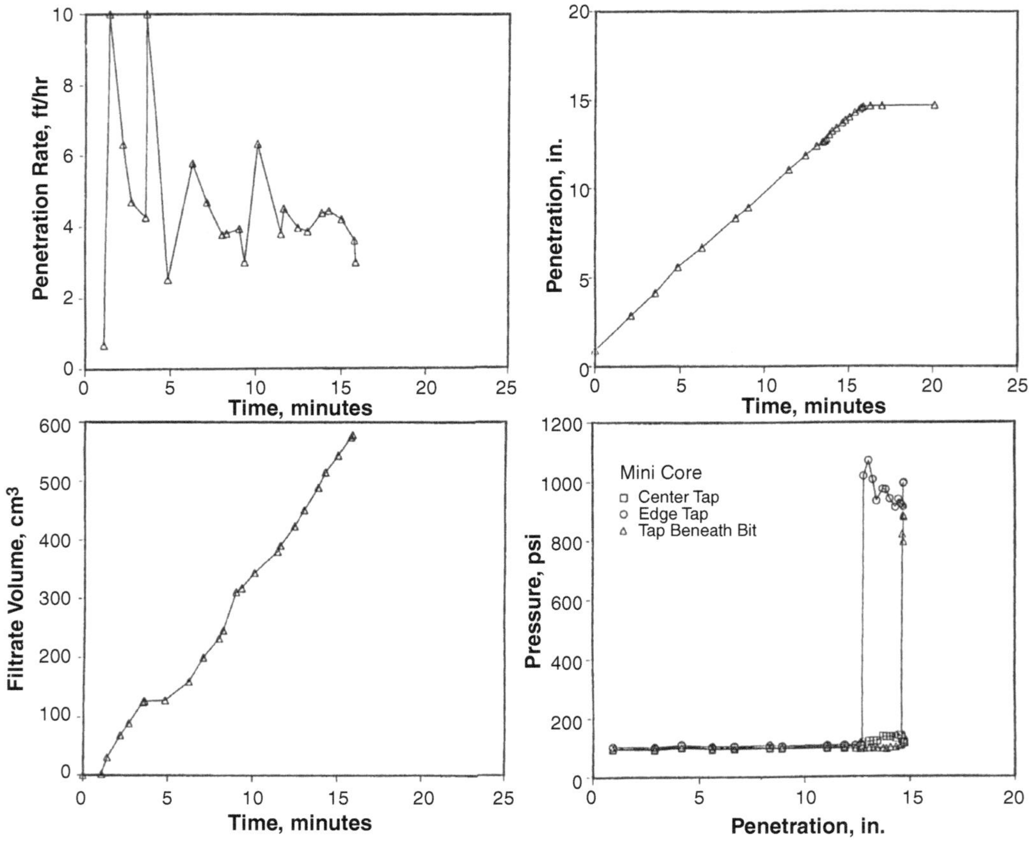

Fig. I.9—DRL Test IV-4 test results. Upper left—Instantaneous coring rate vs. time; upper right—cumulative rock cored vs. time; lower left—cumulative OBM filtrate collected from all four collection port/rings vs. time; and lower right—pressures measured at three pressure taps (see Fig. I.8) vs. penetration. From Warner and Rathmell (1997).

on the difference between the P1 pressure value and the back-pressure kept on the system, the pressure gradient in this increment of rock ahead of the core bit is 1 to 2 psi/in. Finally, the P1 and P2 pressure data (taken after the coring bit is beyond these pressure taps) indicate that the static filter cake on the mini-core has reduced the horizontal pressure gradient within the core to approximately 1 psi/in. These pressure differences are consistent with the concept of an OBM filter cake effectively absorbing almost all of the differential between the drilling mud and the rock pore system. These low-pressure gradients are approximately 10 times the values for reservoir fluid flow, but are unlikely to cause changes in relative permeability because of high capillary number. Capillary numbers in these tests range from 4.3×10^{-5} to 3.5×10^{-4}." *Note: Warner and Rathmell (1997) also cite the Reed et al. (1994) paper discussed earlier.*

The conclusions of Warner and Rathmell (1997) are as follows:

1. "The pressure measurements show that the OBM filter cake totally controls the pressure gradient both in the core and in the rock directly beneath the bit, because almost all of the pressure difference between the mud and formation pressures (even at the cutting face of the core bit) is absorbed across the mudcake.
2. "There will be no movement of connate water, except because of normal water/oil relative permeability effects, because filtrate interfacial tension is only approximately half the reservoir water/oil IFT and pressure gradients are low. Capillary number is below the threshold

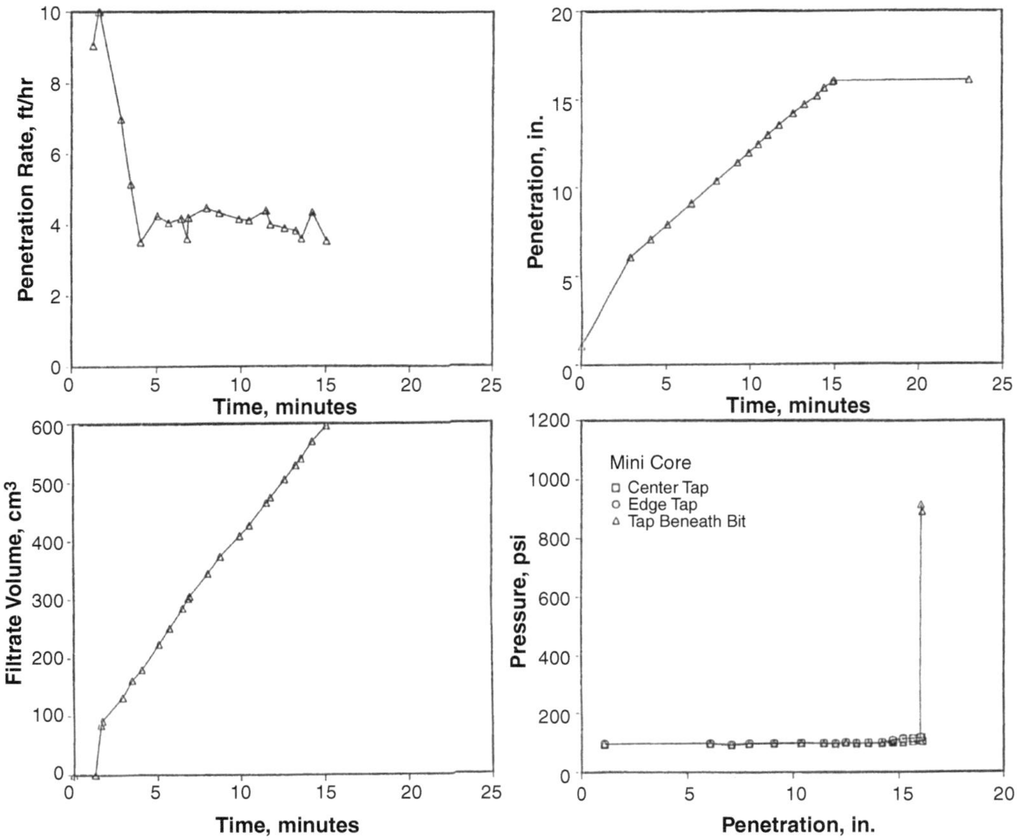

Fig. I.10—DRL Test IV-5 test results. Upper left—Instantaneous coring rate vs. time; upper right—cumulative rock cored vs. time; lower left—cumulative OBM filtrate collected from all four collection port/rings vs. time; and lower right—pressures measured at three pressure taps (see Fig. I.8) vs. penetration. From Warner and Rathmell (1997).

for affecting the relative permeability. For these tests, the invading filtrate properties are similar to reservoir crude when flowing in the core. The filtrate contains only approximately 10% of the carboxylate soap and water found in the whole OBM.

3. "Because of natural variations in S_w within the rock samples, core sample S_w measurements before and after the mini-coring tests could not be used to quantify any potential water movement within the core caused by OBM filtrate invasion.
4. "Above the water/oil transition zone, cores taken with OBM similar to the one used here will yield reliable reservoir S_w values."

I.3 Summary

Mud-filtrate IFT plays an important role in determining if connate water may be mobilized during coring, but, based on capillary number considerations, mudcake is equally important because a high-quality mudcake means very little pressure drop occurs across the actual core. Higher mud-filtrate IFT values also help minimize the possibility of water saturation reduction. The technical studies described herein show that under most circumstances, the IFTs are too large and/or the pressure gradients are too small to result in any connate water mobilization in the cores.

The technical literature contains little information about how reservoir wettability, relative permeability, or electrical properties are actually affected by OBM filtrate and the components and surfactants in the OBM. However, several papers show that under certain conditions, the surfactants typically included in mud formulations can affect rock wettability, which means that rock relative permeability and electrical property measurements can also be affected (Mungan and Moore 1968; Thomas et al. 1984; Sharma and Wunderlich 1985; Ballard and Dawe 1988; Donaldson and Siddiqui 1989; Yan et al. 1993; Sanner and Azar 1994).

Appendix J

Deuterium Oxide Tracer Calculations

A primary goal of low invasion coring (LIC) is to determine the original water saturations in cores so that hydrocarbon reserves can be inferred. Water from coring fluids can interfere with reserves determination. Deuterium oxide (D_2O, or "heavy water") is added to coring fluids to track the invasion of coring fluid filtrate into a core. The D_2O is recovered from cores during Dean-Stark extractions. The D_2O concentrations, in core samples and in the original coring fluid, are determined by a commercial isotope laboratory, and when those data are complete, the following calculations can be used to determine the level of coring fluid invasion.

Note: Salt correction terms in these equations are generally quite small for the reservoir brine:

$$C_{ds}^{d} = C_{mds}^{d} \frac{W_{ds}^{w}}{(W_{ds}^{w} + W^{s})}$$

C_{ds}^{d} = concentration of deuterium in Dean-Stark brine, ppm

C_{mds}^{d} = concentration of deuterium measured in Dean-Stark water, ppm

W^{s} = weight of salt in plug (estimated from produced brine or other core analysis), g

W_{ds}^{w} = weight of water recovered by Dean-Stark extraction, g

Calculate the filtrate saturation in water recovered by Dean-Stark extraction as follows:

$$F = \frac{C_{ds}^{d} - C_{cw}^{d}}{C_{mf}^{d} - C_{cw}^{d}}$$

$$S_{f} = F S_{w}^{ds} \frac{d_{f}^{b}(1 - f_{f}^{s})}{d_{ds}^{b}(1 - f_{ds}^{s})}$$

where

C_{cw}^{d} = concentration of deuterium background in connate water, ppm

C_{ds}^{d} = concentration of deuterium in Dean-Stark brine, ppm

C_{mf}^{d} = concentration of deuterium in mud filtrate, ppm

d_{ds}^{b} = density of Dean-Stark brine, g/cm^3

d_{f}^{b} = density of connate (formation) brine, g/cm^3

F = fraction of filtrate in Dean-Stark brine

f_{ds}^{s} = fraction of salt in the Dean-Stark brine

f_{f}^{s} = fraction of salt in the connate (formation) brine

S_{f} = filtrate saturation, % PV

Calculate the corrected connate water as follows:

$$S_{cw} = S_w^{\,ds} - S_f,$$

where S_w^{ds} = Dean-Stark water saturation, % PV.

Note: If either F exceeds 0.5 or S_f exceeds 10% PV, then do not calculate connate water for this sample.

Record the filtrate saturation and calculated connate water saturation and include these values in the RCAL report.

AUTHOR INDEX

A
Atkins, L.K., 83
Azar, J.J., 112

B
Ballard, T.J., 112
Bell, D.A., 83
Black, A.D., 81–82
Bloys, J.B., 5, 82–83
Bobrosky, D.J., 83
Brownell, L.E., 104–106, 108
Bulling, T.P., 83

C
Charlson, G.S., 5
Collee, P.E., 84

D
Dawe, R.A., 5, 112
Dombrowski, H.S., 104–106, 108
Donaldson, E.C., 112

E
Egbogah, E.O., 5

G
Gale, B.A., 83
Gremley, R.B., 81

H
Holstein, E.D., 5, 82
Huff, C.F., 5
Humphrey, R.A., 82

J
Jenks, L.H., 5
Jerauld, G.R., 83

K
Kralik, J.G., 83
Kriel, W.A., 82

M
McCoy, D.D., 5
Moore, E.J., 112
Mungan, N., 112
Murdock, D.M., 5

P
Parr, J.J., 83
Pasternack, E.S., 83

R
Radford, S.A., 81
Rathmell, J.J., 5, 81–83, 103–104, 106, 108–111
Ray, C.R., 82
Reed, R.L., 82, 103–104, 106–110

S
Sakurai, S., 83
Sanner, D.O., 112
Shallenberger, L.K., 84
Sharma, M.M., 112
Siddiqui, T.K., 112
Skopec, R.A., 84
Spence, A.P., 82
Stalkup, F.I., 104

T
Thomas, D.C., 112
Tibbitts, G.A., 5, 81–83

V
Van Puymbroeck, L., 84

W
Walker, B.H., 81–82
Warner, H.R. Jr., 5, 81–82, 103–106, 108–111
White, E.K., 81
Wilton, B.S., 83
Woodhouse, R., 2
Wunderlich, R.W., 112
Wylie, G.S., 5, 82

Y
Yan, J.-N., 112

SUBJECT INDEX

A
alcohols, 30
American Petroleum Institute (API), 1
  core analysis and, 59
  coring fluids and, 24–25
  crude gravity and, 86
  oil-based mud test and, 91
anaerobic-crude-oil collection letter, 99–100
"Applications of Low Invasion Coring and Outcrop Studies to Reservoir Development Planning for the Villano Field" (Rathmell, Atkins, and Kralik), 83
Atomic Energy Commission, 31–32

B
Baker Hughes Inteq (BHI), 4, 16–20, 39, 45–46, 53–54, 85
barite, 22, 24, 26, 28, 34–36
bentonite, 1, 21–23, 25
biocide, 22–23, 35, 83
bit selection
  BRUTE cutters and, 17
  catcher shoe and, 15
  core diameter and, 17
  coring project manager and, 17
  filter cake and, 16
  lead time and, 17
  oil-based mud (OBM) and, 18–19
  polycrystalline-diamond-compact (PDC) technology and, 4–6, 15, 17, 21, 81, 83
  poorly consolidated sandstone and, 54
  preplanning and, 17–18
  profile and, 15
  project manager and, 17
  rate of penetration (ROP) and, 6, 15, 17–18, 54
  rock type and, 17–18
  rotary speed and, 54
  stabilization and, 15
  throat discharge, 15
  water-based mud (WBM) and, 18–19
  water saturation and, 18
  wettability and, 18
bit whirl, 15
bottomhole assembly (BHA), 40, 44–45, 47
bridging materials, 21, 24, 39
brines, 1, 6
  bromonaphthalene analysis and, 97
  chloride ion analysis and, 8, 101–102
  core analysis and, 66–70
  core barrel and, 19
  coring fluids and, 21, 23, 25–31
  deuterium oxide tracers and, 113
  interfacial tension (IFT) and, 103–111
  iodide ion analysis and, 93
  laboratory operations and, 50
  oil-based mud (OBM) and, 103–109
  original reservoir properties and, 103–111
  preplanning and, 10, 12–13
  selected SPE papers and, 82
  water saturation and, 102
bromide, 28, 30–31, 83
bromonaphthalene analysis
  brines and, 97
  Dean-Stark water extraction and, 97–98
  laboratory and, 95–96, 98
  oil-based mud (OBM) and, 32, 95
  permeability and, 95
  plugging and, 95
  pore volume (PV) and, 95, 98
  rate of penetration (ROP) and, 95
  sampling and, 97
  temperature and, 96–97
  tracers and, 95–98
BRUTE (Backups that are Radically Unaggressive and Tangentially Efficient) cutters, 17
bulk volume, 69

C
calcium carbonate, 8
  barite and, 24
  bentonite and, 23
  coring fluids and, 21–26, 28, 35
  determining amount of in whole mud, 25, 91
  mud weight and, 25
  oil-based mud (OBM) and, 25, 28, 91
  pore-throat-bridging and, 24
  rigsite operations and, 39
  selected SPE papers and, 81, 83
  sized, 21
  viscosity and, 25
  water-based mud (WBM) and, 8, 25, 28, 91
chloride ion analysis, 69
  brines and, 8, 101–102
  Dean-Stark water extraction and, 101–102
  laboratory and, 101
  pore volume (PV) and, 102
  sampling and, 101
  water saturation and, 102
cleaning, 67
collection letter, 99–100
computer tomography (CT) scans, 13, 20, 55, 59–60, 70
contacts list, 13, 71
contingency plans, 13
core analysis
  air permeability and, 67–68
  American Petroleum Institute (API) and, 59
  brines and, 66–70
  bulk volume and, 69
  chloride ion analysis and, 69, 101–102
  cleaning and, 67
  computer tomography (CT) and, 59–60, 70
  core barrel and, 69–70
  core preservation and, 59–66
  coring project manager and, 69, 71
  cutting/trimming plugs and, 62–65
  Dean-Stark water extraction and, 13, 66–67, 71
  drying and, 67
  frozen core and, 70–71
  in-place hydrocarbon volumes and, 60
  laboratory and, 57 (*see also* laboratory)
  lithology data and, 62

material destination and, 69
mean stress for overburden measurements and, 68–69
operator representative and, 59
permeability and, 66–69
plugging and, 59–60, 63, 66, 70
poorly consolidated cores and, 69–71
pore volumes (PV) and, 66–70
porosity and, 67–69, 71
procedures document and, 11–14
routine (RCAL), 1, 3, 13, 62–64, 66, 71, 86
sampling and, 59–60, 62, 69
slabbing and, 13, 59–60, 70–71
special (SCAL), 1, 3, 12–13, 60, 64–66, 71, 86
spectral gamma ray logging and, 59
temperature and, 62, 66, 70
tracers and, 59–60, 66–67, 69, 71
water-based mud (WBM) and, 60
water saturation and, 60, 66–67, 70
wettability and, 60
core barrels, 4, 6
aluminum, 19
bottomhole assembly (BHA) and, 45–46
core analysis and, 69–70
core-catcher assemblies and, 20
full-closure, 55
Gel Coring™ and, 85
high-torque, 19–20
inner, 19, 46
making up, 45–46
mud motors and, 20
outer, 19–20, 45–46
pH and, 19
poorly consolidated sandstone and, 53–56
preplanning and, 12–13, 18
pressure, 7
rigsite operations and, 40, 43–50
second coring run and, 46
selected SPE papers and, 81, 83–84
slabbing and, 20
sponge coring and, 85–86
temperature and, 19–20
tripping in and, 46
tripping out and, 47
core catcher, 4
assemblies, 20
bit selection and, 18
poorly consolidated sandstone and, 54–55
preplanning and, 12
rigsite operations and, 44, 47
shoes for, 12
core-catcher assemblies, 20
core diameter, 1, 4–5
bit selection and, 17
coring fluids and, 21
countercurrent imbibition and, 7
rigsite operations and, 45
Core Labs, 44
core lay-down, 38–40, 44, 47–48
core preservation, 4–5, 23, 25, 84
core analysis and, 59–66
Dean-Stark water extraction and, 66–67
freezing and, 69–71
preplanning and, 11, 13–14
rigsite operations and, 38–39, 50–51
sealant and, 60, 62, 66
sponge coring and, 86
whole-core remnant and, 65–66
wrapping and, 60–62, 65–67
core-processing technicians, 39
core samples. *See* sampling
core sealant, 60, 62, 66
core stabilization, 56–57
coring contractor
equipment and, 39
preplanning and, 10–12
rigsite operations and, 39
coring engineers
bit selection and, 17
core barrels and, 20
poorly consolidated sandstone and, 54, 57
preplanning and, 11–12, 14
rigsite operations and, 38, 45–47
coring fluids
American Petroleum Institute (API) and, 24–25
barite and, 22, 24, 26, 28, 34–36
bentonite and, 23
biocide and, 22–23, 35, 83
brines and, 21, 23, 25–31
calcium carbonate and, 21–26, 28, 35
core diameter and, 21
Dean-Stark water extraction and, 27, 29, 31–32
displacement and, 39
flow rate and, 54
high-pressure/high-temperature (HP/HT), 22–23, 25–26, 28–29
interfacial tension (IFT) and, 25–28
KCl and, 22–25, 35
laboratory and, 21, 26, 28–33
lead time and, 29
lime and, 25–27
low-invasion, 21–36
mud displacements and, 33–35
mud-filtrate tracers and, 29–33
oil-based mud (OBM) and, 6, 21–29, 32–35
permeability and, 4, 21, 24–25, 33, 53
pH and, 22–25, 34
pilot tests and, 22–23, 25–26, 28, 35
plugging and, 33
poorly consolidated sandstone and, 53
pore volumes (PV) and, 29, 31
rate of penetration (ROP) and, 24–26, 28, 33, 35
rigsite operations and, 34, 39
safety and, 29
sampling and, 32, 38–39
shale and, 24–25, 32, 34–35
solids control and, 35–36
specialists for, 14, 38–39
spurt-loss, 53, 81
starch and, 22–23, 25, 35
temperature and, 21–23
tracers and, 4–6 (*see also* tracers)
tripping and, 36
viscosity and, 22–23, 25–26, 28, 33, 35–36
water-based mud (WBM) and, 21–26, 28, 30, 33–35
water saturation and, 22, 25–28
wettability and, 22, 24, 26–27, 35
xanthan gum and, 22–23, 35

coring fluid specialists, 14, 38–39
coring project manager, 6
  bit selection and, 17
  core analysis and, 69, 71
  data collection and, 59
  duties of, 38
  laboratory and, 45
  preplanning and, 9–11
  rigsite operations and, 38, 45, 51
  safety and, 38
  sample collection letter and, 99
  strapping pipe and, 45
"Coring Technology for Point McIntyre Wells" (Wylie, Rathmell, and Tibbitts), 82
countercurrent imbibition, 7, 50, 83

D
Dean-Stark water extraction, 3
  bromonaphthalene analysis and, 97–98
  chloride ion analysis and, 101–102
  core analysis and, 13, 66–67, 71
  coring fluids and, 27, 29, 31–32
  deuterium oxide tracers and, 113–114
  emulsifiers and, 27
  filtrate saturation and, 89–90
  iodide ion analysis and, 93–94
  toluene weight and, 89–90
deuterium oxide tracers
  brines and, 113
  coring fluids and, 29, 31
  Dean-Stark water extraction and, 113–114
  pore volume (PV) and, 113–114
  water saturation and, 113–114
"Development and Application of a Mineral-Oil-Based Mud and Tracer for Low-Invasion Coring at the Point McIntyre Field" (Bloys, Rathmell, Kriel, Spence, Wylie, and Ray), 82
"Development and Application of PDC Core Bits for Downhole Low Invasion Coring in the Arab Carbonates" (Rathmell, Wilton, Gale, Bell, Tibbitts, and Bobrosky), 83
"Development of a Method for Partially Uninvaded Coring in High-Permeability Sandstone" (Rathmell, Tibbitts, Germley, Warner, and White), 81
doglegs, 40
"Downhole Core Preservation for Improved Formation Evaluation" (Skopec, Collee, van Puymbroeck, and Shallenberger), 84
drifting, 45
drilling department, 10
drilling-fluid company, 10
drying, 67
dyes, 30
"Dynamic Spurt-Loss Beneath an Oilfield Bit" (Walker and Black), 81–82

E
effective stress, 68–69
electrical stability (ES), 26, 28–29, 34
electron-capture detector (ECD), 95–97
epoxy, 21, 56–57, 69
equations
  chloride ion analysis, 102
  corrected connate water, 114
  Dean-Stark water extraction, 113–114
  filtrate saturation, 89–90, 113
  mean stress for overburden measurements, 68
  oil saturation, 98
  total weight of oil and contaminants, 97
equipment
  core-analysis contractor, 44–45
  coring contractor, 39–44
  rigsite operations and, 38–45

F
filter cake, 16, 22, 110
filtrate saturation, 113
  Dean-Stark water extraction and, 89–90
  laboratory and, 89–90
  plugging and, 89–90
  pore volume (PV) and, 89
  tracers and, 89–90
flame-ionization detector (FID), 95, 97
flow rate, 54
freezing, 56, 69–71

G
gamma ray scanning, 13, 19, 59–60, 70, 86
gas chromatography (GC), 95–97
Gel Coring^{TM}, 85
Genesis BHC, 54
geologists, 1, 6, 9–10, 38, 49

H
Halliburton, 17, 20
Hamaca field, 54, 81
Hassler sleeve, 68
high-pressure/high-temperature (HP/HT), 22–23, 25–26, 28–29
hydrochloric acid (HCL), 91
Hydro-Lift^{TM}, 38, 44, 48, 53

I
"IFT Flushing Studies to Validate As-Received Oil-Base-Core Water Saturations from the Ivishak (Sadlerochit) Reservoir, Prudhoe Bay Field" (Reed, Humphrey, Holstein, and Rathmell), 82
in-place hydrocarbon volumes, 1, 6, 60
interfacial tension (IFT), 8, 11
  coring fluids and, 25–28
  Dombrowski-Brownell data and, 106, 108
  mud-filtrate relationships and, 103–111
  oil-based mud (OBM) and, 103–111
  original reservoir properties and, 103–111
  select papers and, 82
iodide ion analysis
  brines and, 93
  Dean-Stark water extraction and, 93–94
  laboratory and, 93
  pore volume (PV) and, 94
  sampling and, 93
  tracers and, 31, 93, 93–94
  water saturation and, 93–94
Isotag^{TM}, 32

K
KCl (potassium chloride), 19, 22–25, 35, 66, 70

L

laboratories, 1, 6, 8, 19, 70
  air humidity and, 67, 69
  brines and, 50, 103–111
  bromonaphthalene analysis and, 95–96, 98
  chloride ion analysis and, 101
  contacts and, 71
  core-analysis representative and, 10
  core preservation and, 4–5, 11, 13–14, 23, 25, 38–39, 50–51, 59–66, 84, 86
  coring fluids and, 21, 26, 28–33
  coring project manager and, 45
  crate transport and, 57
  Dean-Stark water extractions and, 67
  field technicians and, 14
  filtrate saturation and, 89–90
  Gel Coring™ and, 85
  iodide ion analysis and, 93
  mud-filtrate/brine IFT relationships and, 103–111
  mud overbalance pressure and, 54
  oil-based mud (OBM) and, 104–108
  operator representative at, 59
  preplanning and, 10–14
  rapid transfer to, 59–60
  rigsite operations and, 39, 49–51, 86
  sample collection letter and, 100
  sampling and, 39
  selected SPE papers and, 82–83
  shipping to, 50–51, 57
  slabbing and, 59–60
  specialized personnel and, 12–14, 57
  sponge coring and, 86–87
  standard procedures of, 59
  tracer analysis and, 33, 69, 89, 113
lead time
  bit selection and, 17
  coring fluids and, 29
  preplanning and, 11–12
  rigsite operations and, 51
lime, 25–27
limestone, 10, 24
lithology data
  core analysis and, 62
  preplanning and, 9, 12
  rigsite operations and, 39
lost-circulation material (LCM), 13, 34, 40
low-invasion coring (LIC), 8
  added benefits of, 5
  bit selection and, 15–18
  bromonaphthalene analysis and, 95–98
  chloride ion analysis and, 101–102
  core analysis and, 59–71 (*see also* core analysis)
  core barrels and, 19–20
  defined, 1
  filtrate saturation and, 89–90
  Gel Coring™ and, 85
  higher cost of, 6, 9
  historical perspective on, 5
  iodide ion analysis and, 93–94
  key aspects of, 4–5
  limitations of, 6–7
  objectives of, 3–4
  oriented coring and, 10
  overview of, 1–2
  poorly consolidated sandstone and, 53–57
  preplanning and, 9–14
  questions to ask about need for, 9–10
  rigsite operations and, 37–51
  sample collection letter and, 99–100
  selected SPE papers on, 81–84
  sponge coring and, 85–87
  tracers and, 113–114 (*see also* tracers)
  water saturation and, 103–112 (*see also* water saturation)
"Low Invasion, Synthetic Oil-Base Mud Coring in the Yacheng 13-1 Gas Reservoir for Gas-in-Place Calculation" (Rathmell, Bloys, Bulling, Parr, Pasternack, Jerauld, and Sakurai), 83
low-shear-rate viscosity (LSRV), 35

M

materials/logistics department, 6, 10
mean stress, 68–69
"Mechanisms Controlling Filtration at the Core Bit for Oil-Based Muds" (Warner and Rathmell), 82
miscible gas, 9
mobile water, 7, 85
mud motors, 20
mud overbalance pressure, 54

N

"New Technology and Tools for the Recovery of Representative Cores from Uncemented Sand Formations" (Tibbitts and Radford), 81
nitrogen gas, 60, 63–64, 69
North Slope, 30

O

Oil-based mud (OBM), 1, 5
  adapting for coring, 28–29
  adequate lead time and, 11
  bentonite and, 21
  bit selection and, 18–19
  bromonaphthalene analysis and, 32, 95
  calcium carbonate and, 25, 28, 91
  controlling water contamination in, 27–28
  coring fluids and, 6, 21–29, 32–35
  countercurrent imbibition and, 7
  electrical stability (ES) and, 26, 28–29, 34
  formation brine confusion and, 27
  high-pressure/high-temperature (HP/HT), 23, 28
  laboratory and, 104–108
  metal cans and, 33
  mud-filtrate/brine IFT relationships and, 103–111
  oil-based coring fluid and, 35
  original reservoir properties and, 103–111
  permeability and, 103–112
  preplanning and, 11
  rate of penetration and, 18, 28
  rigsite operations and, 47
  selected SPE papers on, 81–83
  sponge coring and, 85
  spurt losses and, 21
  superiority of, 39
  surfactants and, 28, 103, 112
  temperature and, 83, 106
  tracers and, 28, 32
  tripping out and, 47

very hard rocks and, 7
water-based coring fluid and, 35
water contamination and, 27–28
water saturation and, 25–26, 103–112
wettability and, 103–104, 106, 111–112
oil/water contact (OWC), 7, 82
operator core-analysis specialist, 14, 59, 71
oriented coring, 10
overburden measurements, 68–69

P

permeability, 3, 5
bromonaphthalene analysis and, 95
core analysis and, 66–69
coring fluids and, 4, 21, 24–25, 33, 53
oil-based mud (OBM) and, 103–112
preplanning and, 9, 12–13
rigsite operations and, 39
selected SPE papers and, 81, 83–84
tripping rates and, 55
water saturation and, 95 (*see also* water saturation)
pH, 6
aluminum inner core barrels and, 19
coring fluids and, 22–25, 34
very hard rocks and, 7
photography, 5, 13, 69–71, 86
pilot tests
coring fluids and, 22–23, 25–26, 28, 35
preplanning and, 11, 14
rigsite operations and, 38
plaster of Paris, 56–57, 69
plugs, 4
bromonaphthalene analysis and, 95–98
core analysis and, 59–60, 63, 66, 70
coring fluids and, 33
cutting/trimming plugs and, 62–65
filtrate saturation and, 89–90
Gel Coring™ and, 85
poorly consolidated sandstone and, 55
preplanning and, 13
rigsite operations and, 38–39, 45, 50–51
Point McIntyre, 82, 90
polyanionic cellulose (PAC), 22–23, 25, 35
polycrystalline-diamond-compact (PDC) bits, 4–6, 15, 17, 21, 81, 83
poorly consolidated cores, 69–71
poorly consolidated sandstone
bit selection and, 54
core catcher and, 54–55
core diameter and, 53
core length and, 53
core stabilization and, 56–57
coring fluids and, 53
flow rate and, 54
full-closure core barrel and, 55
mud overbalance pressure and, 54
plaster of Paris and, 57
pore volumes (PV) and, 55
shale and, 54–55
temperature and, 55, 57
tripping rates and, 55–56
Pore volume (PV). *See also* water saturation
bromonaphthalene analysis and, 95, 98
chloride ion analysis and, 102
core analysis and, 66–70
coring fluids and, 29, 31
deuterium oxide tracers and, 113–114
filtrate saturation and, 89
Gel Coring™ and, 85
iodide ion analysis and, 94
low-displaceable, 7
poorly consolidated sandstone and, 55
sponge coring and, 86–87
tracers and, 29
tripping rates and, 55
porosity, 3
core analysis and, 67–69, 71
low-displacement pore volume and, 7
preplanning and, 9, 12–13
preplanning
adequate lead time and, 11–12
bit selection and, 17–18
brines and, 10, 12–13
contact list and, 13
contingency plans and, 13
core-analysis procedures document and, 11–14
core barrel and, 12–13, 18
coring project manager and, 9–11
formation temperature and, 10
geologists and, 9
laboratory and, 10–14
lead time and, 11–12
lithology data and, 9, 12
objectives of, 9–10
oil-based mud (OBM) and, 11
permeability and, 9, 12–13
pilot tests and, 11, 14
plugging and, 13
porosity and, 9, 12–13
project manager and, 9–12, 14, 17
questions to ask, 9–11
rig supervisor and, 10, 12
routine core analysis (RCAL) and, 13
special core analysis (SCAL) and, 12–13
specialized personnel and, 12–13
team approach and, 10–11
tracers and, 12–13
water-based mud (WBM) and, 9
water saturation and, 9, 12–13
project manager, 6
bit selection and, 17
duties of, 38
preplanning and, 9–12, 14, 17

Q

quality-control/quality-assurance (QC/QA)
core analysis and, 59
preplanning and, 12–14
rigsite operations and, 44–45, 50

R

rate of penetration (ROP)
bit selection and, 6, 15, 17–18, 54
bromonaphthalene analysis and, 95
coring fluids and, 24–26, 28, 33, 35
flow rate and, 54
mud motors and, 20
oil-based mud (OBM) and, 18, 28

preplanning and, 10, 13–14
rigsite operations and, 38–40, 46–47
rotary speed and, 54
selected SPE papers and, 81–84
Recommended Practices for Core Analysis (API), 1
reservoir engineer, 6, 9–11
residual oil saturation, 5, 9, 103
rigsite operations
bottomhole assembly (BHA) and, 40, 44–45, 47
core-analysis procedures document and, 12–13
core barrel and, 40, 43–50
core catcher and, 44, 47
core diameter and, 45
core lay-down and, 38–40, 44, 47–48
core processing and, 39, 49–50
coring contractor and, 39
coring engineers and, 38
coring-fluid specialist and, 38–39
coring mechanics and, 46–47
coring project manager and, 38, 45, 51
drifting pipe and, 45
equipment and, 38, 39–45
geologists and, 38, 49
laboratory and, 39, 49–51, 86
lead time and, 51
lithology data and, 39
making up core barrel and, 45–46
oil-based mud (OBM) and, 47
overview of, 37
permeability and, 39
personnel and, 37–39
pilot tests and, 38
plugging and, 38–39, 45, 50–51
quality-control and, 44–45, 50
rate of penetration (ROP) and, 38–40, 46–47
roustabouts and, 39
safety and, 12, 38–39, 44–45, 47–48
sampling and, 38–39, 50, 56–57
shale and, 39, 46
slabbing and, 38–39, 50
specialized personnel and, 12–13
sponge coring and, 86
strapping pipe and, 44–45
temperature and, 46, 51
tracers and, 38–39, 49–50
tripping and, 37, 46–47
water-based mud (WBM) and, 39
water saturation and, 50
wettability and, 39, 49–50
rig supervisor, 6
coring fluids and, 34
lead time and, 51
preplanning and, 10, 12
rigsite geologist and, 38
role of, 38
rotary speed, 54
roustabouts, 39
routine core analysis (RCAL)
cutting and trimming core plugs and, 62–64
Dean-Stark water extraction and, 3 (*see also* Dean-Stark water extraction)
laboratory operations and, 71
lengthy documents for, 1
oil permeability and, 66
operator point contact and, 71
preplanning and, 13
vertical plug location and, 63–64

S

safety
core lay-down and, 47–48
coring fluids and, 29
coring project manager and, 38
equipment for, 44–45
information for, 29
rigsite operations and, 12, 38–39, 44–45, 47–48
training for, 39
tripping out and, 47
sampling, 1, 3
brine saturation and, 101
bromonaphthalene analysis and, 95, 97, 113
chloride ion analysis and, 101
cleaning and, 67
collection letter and, 99–100
core analysis and, 59–60, 62, 69
coring fluids and, 32
countercurrent imbibition and, 7
drying and, 67
Gel Coring™ and, 85
iodide ion analysis and, 93–94
laboratory operations and, 86
preplanning and, 13–14
preservation and, 4–5, 11, 13–14, 23, 25, 38–39, 50–51, 59–66, 84, 86
quality-control and, 50
rigsite operations and, 38–39, 50, 56–57
routing program for, 60
salt crystals and, 67
sponge coring and, 85–87
water saturation and, 111
sandstone
Bentheimer, 21
Berea, 21, 67
core barrel and, 53–56
gold, 21
poorly consolidated, 53–57
scanning electron microscopy (SEM), 13
shale, 10, 20
coring fluids and, 24–25, 32, 34–35
coring poorly consolidated sandstone and, 54–55
rigsite operations and, 39, 46
shucks, 40, 41–42, 47–49
slabbing
core analysis and, 13, 59–60, 70–71
core barrel and, 20
Gel Coring™ and, 85
rigsite operations and, 38–39, 50
sponge coring and, 86
solids control, 35–36
South China Sea, 19
spacers, 33–35, 39, 50, 82
special core analysis (SCAL), 3
cutting/trimming plugs and, 64–65
future plug preservation and, 60
homogeneous rock and, 71
laboratory operations and, 86
lengthy documents for, 1
plug preservation and, 65–66

preplanning and, 12–13
whole-core remnant and, 65–66
specialized personnel, 12–13
sponge coring
core barrel and, 85–86
laboratory and, 86–87
oil-based mud (OBM) and, 85
pore volume (PV) and, 86–87
rigsite operations and, 86
sampling and, 85–87
slabbing and, 86
tracers and, 86
spurt-loss tests, 11, 23, 81
starch, 22–23, 25, 35
strapping, 44–45, 59, 62
surfactants
core barrels and, 19
coring fluids and, 22, 26–28, 33, 35
oil-based mud (OBM) and, 28, 103, 112
very hard rocks and, 7

T

team approach, 10–11
temperature, 7, 100
bromonaphthalene analysis and, 96–97
core analysis and, 62, 66, 70
core barrels and, 19–20
coring fluids and, 21–23
oil-based mud (OBM) and, 83, 106
poorly consolidated sandstone and, 55, 57
preplanning and, 10
rigsite operations and, 46, 51
toluene, 32, 66–67, 89–90, 95–98
total flow area (TFA), 12, 39
tracers, 8
adding to mud system, 30
alcohols and, 30
background measurement and, 32–33
bromide, 28, 30–31, 83
bromonaphthalene analysis and, 95–98
characteristics of good, 29
core analysis and, 59–60, 66–67, 69, 71
coring fluids and, 4–6, 22, 24, 26–33, 36
cost of, 33
countercurrent imbibition and, 7
deuterium oxide, 29, 31, 113–114
dyes and, 30
filtrate saturation and, 89–90
iodide ion analysis and, 31, 93–94
Isotag™ and, 32
mud-filtrate, 29
oil-based mud (OBM) and, 32
preplanning and, 12–13
rigsite operations and, 38–39, 49–50
select papers and, 82–83
sponge coring and, 86
tritium, 31–32
water-based mud (WBM) and, 30–32
water phase and, 28
tripping, 5
coring fluids and, 36
in, 46
out, 47
poorly consolidated sandstone and, 55–56
rates of, 47, 55–56
rigsite operations and, 37, 46–47
tritium, 31–32, 66

U

US Department of Transportation (DOT), 33, 67
UV light, 69, 71, 86–87

V

Venezuela, 54
very hard rocks, 7
viscosity, 7
calcium carbonate and, 25
coring fluids and, 22–23, 25–26, 28, 33, 35–36
low-shear-rate (LSRV), 35
oil-based mud (OBM) and, 104
plaster of Paris and, 57
polypropylene glycol gel and, 84
tripping rates and, 55

W

water-based mud (WBM), 1
bentonite and, 21
biocides and, 22
bit selection and, 18–19
calcium carbonate and, 8, 25, 28, 91
core analysis and, 60
coring fluids and, 4, 21–22, 25–26, 28, 30, 33–35
countercurrent imbibition and, 7
fluid-loss polymers and, 21
high-pressure/high-temperature fluid loss and, 26
historical perspective on low-invasion coring (LIC) and, 5
oil-based coring fluid and, 34
preplanning and, 9
relative permeability and, 39
residual oil saturation and, 5, 9
rigsite operations and, 39
selected SPE papers on, 81–83
spacers and, 34–35
spurt losses and, 21
tracers and, 30–33
tripping in and, 46
very hard rocks and, 7
water-based coring fluid and, 34–35
water saturation and, 60
wettability and, 39
water saturation, 1, 6
bit selection and, 18
brine and, 102
chloride ion analysis and, 102
core analysis and, 60, 66–67, 70
coring fluids and, 22, 25–28
deuterium oxide tracers and, 113–114
impact of oil-based mud on, 102–112
iodide ion analysis and, 93–94
mud-filtrate/brine IFT relationships and, 103–111
oil-based mud (OBM) and, 25–26, 103–112
original reservoir properties and, 103–111
preplanning and, 9, 12–13
rigsite operations and, 50
selected LIC papers and, 82–84
water-based mud (WBM) and, 60
wellsite. *See* rigsite operations

wettability, 1, 3
bit selection and, 18
core analysis and, 60
coring fluids and, 22, 24, 26–27, 35
in-situ, 18
oil-based mud (OBM) and, 103–104, 106, 111–112
rigsite operations and, 39, 49–50
very hard rocks and, 7

X

xanthan gum, 22–23, 35
X-ray diffraction (XRD), 13, 70

www.ingramcontent.com/pod-product-compliance
Ingram Content Group UK Ltd.
Pitfield, Milton Keynes, MK11 3LW, UK
UKHW060010051026
15591UKWH00002B/138
* 9 7 8 1 5 5 5 6 3 3 0 3 5 *